Essentials of UWB

If you are involved in designing, building, selling, or regulating UWB devices, this concise and practical guide to UWB technology, standards, regulation, and intellectual property issues will quickly bring you up-to-speed. Packed with practical insights, implementation guidelines and application examples, *Essentials of UWB* is a must-have resource for wireless professionals working in the field.

Written by key figures in the development of UWB, the book describes UWB technology, and evaluates its suitability for applications in communications, radar, and imaging. UWB radios, protocols, and implementation are covered, and a thorough account of UWB industry organization completes the picture.

This is an invaluable guide for engineers involved in UWB device design, as well as for product-marketing managers, sales-support engineers, and technical managers. It will also appeal to engineers with a deeper technical understanding of UWB who want to gain knowledge of the broader environment and future evolutionary expectations.

STEPHEN WOOD is a Technology Strategist at Intel Corporation. He was one of the original founders of the OFDM Alliance and has been President of WiMedia for the last three years.

ROBERTO AIELLO is Founder and Chief Technical Officer of Staccato Communications, and former Founder, President, and CEO of Fantasma Networks. A recognized leader in the UWB community, Dr Aiello built the first documented UWB network at Intervel Research, Paul Allen's Research Laboratory.

The Cambridge Wireless Essentials Series

Series Editors
WILLIAM WEBB, *Ofcom, UK*
SUDHIR DIXIT, *Nokia, US*

A series of concise, practical guides for wireless industry professionals.

Martin Cave, Chris Doyle and William Webb, *Essentials of Modern Spectrum Management*
Christopher Haslett, *Essentials of Radio Wave Propagation*
Stephen Wood and Roberto Aiello, *Essentials of UWB*

Forthcoming
Chris Cox, *Essentials of UMTS*
Steve Methley, *Essentials of Wireless Mesh Networking*
Linda Doyle, *Essentials of Cognitive Radio*
David Crawford, *Essentials of Mobile Television*
Malcolm Macleod and Ian Proudler, *Essentials of Smart Antennas and MIMO*

For further information on any of these titles, the series itself and ordering information see www.cambridge.org/wirelessessentials.

Essentials of UWB

Stephen Wood
Intel Corporation

Dr Roberto Aiello
Staccato Communications

Shaftesbury Road, Cambridge CB2 8EA, United Kingdom

One Liberty Plaza, 20th Floor, New York, NY 10006, USA

477 Williamstown Road, Port Melbourne, VIC 3207, Australia

314–321, 3rd Floor, Plot 3, Splendor Forum, Jasola District Centre, New Delhi – 110025, India

103 Penang Road, #05–06/07, Visioncrest Commercial, Singapore 238467

Cambridge University Press is part of Cambridge University Press & Assessment, a department of the University of Cambridge.

We share the University's mission to contribute to society through the pursuit of education, learning and research at the highest international levels of excellence.

www.cambridge.org
Information on this title: www.cambridge.org/9780521877831

First published 2008

A catalogue record for this publication is available from the British Library

Library of Congress Cataloging-in-Publication data
Wood, Stephen R.
Essentials of UWB/ Stephen Wood, Dr Roberto Aiello.
p. cm.
Includes bibliographical references and index.
ISBN 978-0-521-87783-1 (hbk.)
1. Broadband communication systems. 2. Ultra-wideband devices.
3. Wireless communication systems. I. Aiello, Roberto. II. Title.
III. Title: Essentials of ultra-wideband.
TK5103.4.W66 2008
621.384–dc22 2008006278

ISBN 978-0-521-87783-1 Hardback

Contents

1	Introducing ultra-wideband (UWB)	*page* 1
	1.1 Ultra-wideband application classes	3
	1.1.1 High-data-rate communications	3
	1.1.2 Low-data-rate communications	6
	1.1.3 Imaging	7
	1.1.4 Automotive radar	7
	1.2 Next-generation HDR applications	8
	1.3 A brief history of ultra-wideband	11
	1.4 Summary	17
2	Matching UWB to HDR applications	19
	2.1 Speed – specifying UWB	19
	2.2 Low cost	26
	2.3 Location	28
	2.4 Low power consumption	30
	2.5 Personal area network architecture	31
	2.5.1 Range does not equal goodness	31
	2.5.2 The natural stratification of wireless networks	32
	2.6 Summary	35
3	Physical-layer (PHY) characteristics	37
	3.1 Multiband	41
	3.2 Multiband orthogonal frequency-division multiplexing	43
	3.3 Summary	46
4	Media-access control (MAC) layer	48
	4.1 Channel selection	52
	4.2 Beaconing and synchronization	53

4.3 Multi-rate support 55
4.4 Transmit-power control (TPC) 55
4.5 Power management 56
4.6 Range measurement 57
4.7 Bandwidth reservations 58
4.7.1 Prioritized contention access (PCA) 58
4.7.2 Distributed reservation protocol (DRP) 60
4.8 Co-existence of different protocols 61
4.9 Wireless USB MAC functions 62
4.9.1 Wireless USB addressing 63
4.9.2 Host channel 64
4.10 Summary 64

5 Implementation information 66
5.1 Co-location with other radios on the same platform 66
5.2 Chip-integration considerations 70
5.2.1 Integration 70
5.2.2 Packaging 72
5.3 Antenna considerations 73
5.3.1 Antenna types 74
5.3.2 Antenna requirements 75
5.3.3 Antenna availability 75
5.4 Radios built on cards vs. integrated designs 76
5.5 Summary 79

6 Upper-layer protocols 81
6.1 Certified wireless USB (CWUSB) 82
6.1.1 Main applications of CWUSB 83
6.1.2 System architecture 83
6.1.3 Protocol description 84
6.1.4 Strengths and weaknesses 85
6.1.5 Main challenges 85
6.1.6 Application example 86
6.2 WiMedia layer-two protocol (WLP) 86
6.2.1 Main applications 87
6.2.2 System architecture 88

6.2.3 *Protocol description* 89
6.2.4 *Main challenges* 90
6.2.5 *Strengths and weaknesses* 91
6.2.6 *Application example* 91
6.3 Bluetooth 92
6.3.1 *Main applications* 93
6.3.2 *System architecture* 93
6.3.3 *Protocol description* 95
6.3.4 *Main challenges* 95
6.3.5 *Strengths and weaknesses* 96
6.3.6 *Application example* 97
6.4 Wireless 1394 97
6.4.1 *Main applications* 97
6.4.2 *System architecture* 97
6.4.3 *Protocol description* 98
6.4.4 *Main challenges* 98
6.4.5 *Strengths and weaknesses* 98
6.4.6 *Application example* 98
6.5 Association 99
6.6 Summary 101

7 Ultra-wideband standardization 103
7.1 Ecma International 103
7.2 International Standards Organization (ISO) 104
7.3 ETSI 105
7.4 An international perspective on standardization 106
7.5 Standards' role in international trade 108
7.6 Ultra-wideband in the IEEE 110
7.7 Summary 114

8 Special-interest groups 115
8.1 An overview of UWB special-interest groups 116
8.2 The WiMedia Alliance 117
8.3 The Bluetooth SIG 120
8.4 Universal-Serial-Bus Implementer's Forum 121
8.5 Other SIGs 124

8.6 Special-interest-group (SIG) operations relating to UWB 124
8.6.1 Intellectual-property rights 125
8.6.2 Interoperability and certification testing 131
8.6.3 Membership rights in SIGs 132
8.7 Summary 135

9 Ultra-wideband business issues 137
9.1 Expected changes to the technology over time 137
9.1.1 Planned development in UWB 137
9.1.2 Multiple-radio integration 139
9.1.3 Converging the WAN, LAN and PAN networks 139
9.2 Business and market trends 141
9.2.1 Price erosion 142
9.2.2 Consolidation 143
9.2.3 Rollout expectations 144
9.3 Summary 146

10 Regulating ultra-wideband 148
10.1 An overview on regulation 149
10.2 The beginnings of UWB regulation 150
10.3 Protection vs. innovation 151
10.4 European regulatory leadership 152
10.5 European regulatory bodies and organizations 153
10.5.1 The national administration 154
10.5.2 CEPT 154
10.5.3 The European Union, European Commission and Radio Spectrum Committee 155
10.6 The challenge of regulating UWB in Europe 157
10.7 The first mandate – technical work begins 158
10.7.1 Characterizing ultra-wideband 159
10.7.2 Evaluating UWB's interference potential 160
10.7.3 A zero-interference assumption 163
10.7.4 Report 64 163

10.8 The second mandate 165
10.9 The third mandate 168
10.10 Single entry vs. aggregation 170
10.11 The need for ongoing regulatory work 174
10.12 Moving above 6 GHz 175
10.13 Mitigation techniques 177
10.13.1 Low-data-rate communications (LDC) 178
10.13.2 Detect and avoid (DAA) 179
10.13.3 Ten-second rule 182
10.13.4 No outdoor infrastructure 183
10.13.5 Mains attached 184
10.14 Summary 184

11 Tragedy of the commons 187
11.1 Ultra-wideband spectrum saturation 187
11.2 Saturation of WLAN due to PAN applications 189
11.3 Summary 192

Appendix: Reference documents 194
Author biographies 200
Index 201

1 Introducing ultra-wideband (UWB)

If you are interested in a deep theoretical treatise on ultra-wideband, there are several excellent texts, which are listed at the end of this chapter, that we recommend [1, 2]. *Essentials of UWB* will definitely not fill that need. It is far too concise and practical and it fails to take up the requisite three inches of shelf space that are required to fill that niche in the literature.

If you are an engineer, business professional, regulator or marketing person who needs enough technical information to build, sell or regulate products that include a UWB radio, but don't aspire to become a radio frequency (RF) deity in your own right, this is the text that you are looking for. Our objective in writing this book is to provide a dependable overview of the data that you need to know to understand the technology and the industry. This includes technical overviews, industry organization, intellectual property overview, standardization and regulatory discussions. We will also attempt to provide pointers to source documents for deeper investigation for those who are so inclined. We know where the good data are buried because in many cases we had a hand in putting it there. Dr Aiello founded two UWB start-ups, contributed actively to the US regulatory processes, participated in the IEEE standardization wars and performed much of the early development of UWB modulation schemes and radio designs. He has also been a board member in the WiMedia Alliance for a number of years. Mr Wood has been the President of the WiMedia Alliance for several years, has participated in regulatory proceedings in the ITU (International Telecommunications Union) and in Europe, has been a principal architect for the industry structure, has been heavily involved in WiMedia's relationship with Ecma, ISO and ETSI and has played a major role in the development of WiMedia's intellectual-property strategy.

So, with the preliminaries out of the way, it is only appropriate to begin with a description of ultra-wideband. Describing UWB is slightly more difficult than one would immediately suspect. While the press may speak of UWB as a single technology, it is not. Ultra-wideband is best described as a class of radios that use unusually wide bandwidth signals to achieve their application objectives. These radios can operate over a wide range of frequencies, and with different signal characteristics. By way of example to demonstrate the variability in UWB: automotive UWB radars are allowed to operate in the 24 GHz band and require licences; US data communication applications operate over a 3.1–10.6 GHz range and are unlicensed; ground-penetrating radars in the USA operate over the same frequencies as data-communication devices and at the same power levels, but require licensing and have additional restrictions about their use; and the same data-communication applications that operate over 3.1–10.6 GHz in the USA are allowed to operate in a more limited range of frequencies in other parts of the world and with additional restrictions to reduce the probability of generating harmful interference. As one can see, for a single technology, there are very few common elements that can readily be used to define it.

The principal point that is common is that relatively low power is being traded off against wide bandwidths to obtain enough performance to be economically interesting. The very low power levels of UWB would appear to be a major shortcoming, but instead they are offset by the exceedingly wide available bandwidth, which allows the technology to operate at extremely high data rates. Ultra-wideband is distinct from most other commercial technologies because of its underlay philosophy. Figure 1.1 demonstrates the very wideband nature of UWB as compared with more traditional narrowband signals.

Before UWB, the spectrum was divided up by frequency (principally) and only a very restricted overlap of authorized services was allowed. By contrast, UWB is intentionally designed to overlap a broad swath of other services, as is depicted in Figure 1.1. The regulators who initially authorized UWB believe that a higher-priority service would tend to overpower UWB when it needs the spectrum and that UWB would have additional frequencies available that would allow it to move out of the way when this occurs. By implementing an underlay, regulators were attempting to increase the amount of spectrum that was

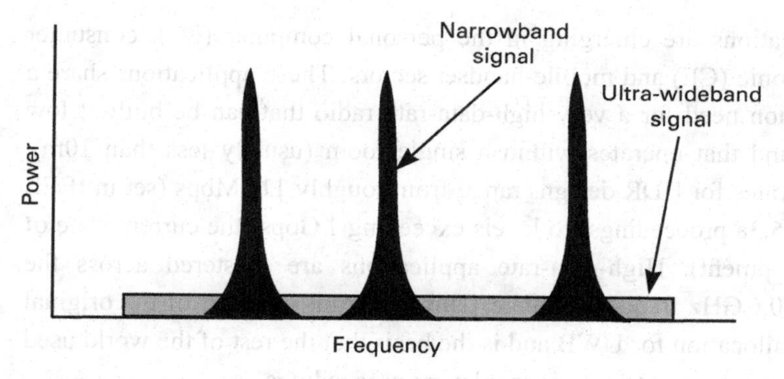

Figure 1.1 Narrowband vs. wideband signals

available for an increasing volume of communication needs. This logic has since been proven to be somewhat less than perfect, but it was the prevailing opinion when the Federal Communication Commission (FCC) made its first rulings in 2002.

1.1 Ultra-wideband application classes

With this description of UWB in hand, the best way to proceed is to provide an abbreviated overview of the applications that employ UWB. If one were to comb through the FCC documents as well as those generated in the ITU and European and Asian proceedings, one would find a more exhaustive list than is being provided here, but also one that tends toward the esoteric. The more practical applications are first grouped into four broad categories including high-data-rate communications, low-data-rate communications, general imaging and automotive radar. The applications are grouped this way because the regulatory proceedings have been crafted around these groupings and will figure in the discussion on regulation during Chapter 8. This structure is also reflective of the market. There are very separate market sectors promoting these products and developing standards.

1.1.1 High-data-rate communications

High-data-rate (HDR) communication is the first application class and the one which will be the principal focus of this text. High-data-rate applications are emerging in the personal computer (PC), consumer electronic (CE) and mobile-handset sectors. These applications share a common need for a very high-data-rate radio that can be built at low cost and that operates within a single room (usually less than 10m). Data rates for HDR designs range from roughly 110 Mbps (set in IEEE 802.15.3a proceedings) to levels exceeding 1 Gbps (the current state of development). High-data-rate applications are clustered across the 3.1–10.6 GHz frequency range. This spectrum was part of the original FCC allocation for UWB and is the basis that the rest of the world used as a starting point in their regulatory proceedings.

To be a little more specific, HDR applications can be broken down to include file transfer, asynchronous communications, streaming video and streaming audio. File transfers may be an exchange that is point-to-point, such as moving music files from two mobile players when schoolchildren wish to share music. Likewise, transferring pictures from a digital camera to a printer, loading movie files to a portable video player or downloading a game from a kiosk would all be considered point-to-point applications.

A point-to-point application is a special-purpose exchange. There are only two parties in the communication. The link is set up exclusively to perform the transaction and is usually taken down when the transaction is concluded. This is a rather special-purpose approach to communication. By contrast, networked communications tend to be a more generalized superset. Instead of assuming that only two devices are active, they are provisioned to be able to handle multiple concurrent conversations across a shared media.

A computer sending a word-processing document to a printer would be an example of a point-to-point exchange, but one that would be likely to be transacted across the more general networked environment. This would allow a scanner or a remote hard drive to share the link with the printer instead of having separate purpose-built radio. Because of the more versatile nature of a networked architecture, the designs incorporated in all HDR implementations deployed today use that approach.

The second group of HDR applications is asynchronous communications. The term 'asynchronous communications' is intended to describe an ongoing, intermittent stream of exchange or conversation. A wireless keyboard or monitor would be an example. In asynchronous communications, the link may remain persistent, and the exchanges are generally thought of less as files and more as individual data blocks.

The third group of HDR applications is streaming video. In this group, a continuous stream of video data is carried by the radio. Usually, this involves feeding a video display or transferring video between two devices for eventual display. Streaming has a normal assumption that the data are being consumed in real time by a person. The term streaming implies that the data are being fed at a consistent rate, which is gauged to be the normal rate of consumption by a person. Streaming video may be moved between a DVD or a game console and a television. As a general rule, streaming video is moved between storage devices and displays. If a human being is not in the direct path, it is frequently more effective to perform a file transfer instead.

Like streaming video, streaming audio has the same assumption of being transferred at a data rate synchronized to the consumption rates that people expect. Unlike streaming video, streaming audio operates at much slower data rates. Ultra-wideband is frequently overkill for this traffic load and would generally only be employed if higher fidelity audio quality is required. Audio distribution also requires synchronization and quality of service. Streaming audio would also be used for wireless speakers in a stereo system or in a home theatre.

Within the high-data-rate category of applications, there will be a noticeable evolutionary shift, which will take place over two generations. The first generation of UWB applications will be all about removal of the cable and replacing it with a high-speed wireless connection. Customers are interested in doing this principally to get rid of the cable mess that hides behind most PCs and televisions. By itself, this has the potential to be a market of satisfactory size and value. It would be reasonable to say that this is what UWB will become. But this isn't the end of the story.

In the second generation, UWB has the potential to become a short-range, high-speed wireless infrastructure, which connects mobile platforms to stationary platforms and to each other. It enables a wide range of new-use models that were not practical without it. Just as the emergence of the Ethernet enabled email by connecting PCs together, UWB performs the same function for mobile devices. In the PC world, dial-up lines and RS232 cables provided connectivity, but not under terms which were sufficient for the application to operate effectively. In the mobile environment, the WAN (wide area network) and low-rate Bluetooth connections are equally inefficient at connecting the mobile platform to various stationary exchange partners.

As an example, there is a great deal of discussion in the consumer electronics community around the deployment of kiosk networks, which will allow mobile devices to engage a kiosk at a grocery store or train station in order to purchase MP3 files, maps, electronic postcards or other services. Evolving UWB technology could allow you to put the moral equivalent of a kiosk connection into your car or home. Ultra-wideband connections could provide the common touch point that allows mobile platforms to transfer data with stationary systems and the Internet.

1.1.2 Low-data-rate communications

The second major application class is that of low-data-rate communications (LDR). Low-data-rate applications are synonymous with sensor networks for all practical purposes. Inexpensive, low-power sensors are deployed within a building, in a factory, in agricultural fields and a variety of other places. Sensor networks are used for intelligent lighting and energy control within buildings, factory automation and warehousing applications. Typically, these applications involve the transfer of very small volumes of data between battery-powered transceivers. Some applications have a requirement for tracking the physical location of goods and use the unique characteristics of the UWB signal to establish a precise physical location of the transmitter. Sensor networks trade off peak throughput in favour of extended range. The increased range and link margin that they gain by this trade-off allows them to operate a network with low infrastructure costs using very small batteries.

Unlike the HDR market, which uses UWB almost exclusively, there are a number of proprietary UWB radio designs that are employed for LDR applications. While the total sensor market is expected to be quite large, the total market volume will be split between these various radio technologies. Because of this it is unclear at this time whether the UWB contribution to the LDR market will be particularly significant.

1.1.3 Imaging

The third major application category is general imaging. Imaging is a category that includes ground-penetrating radar (GPR), through-wall imaging, in-wall imaging and security perimeters. Ground-penetrating radar is used by utility companies, construction companies and archeologists to look for objects that are below the surface of the ground. Through-wall imaging is used by police and the military to look into adjacent rooms for people, obstacles and hazards. In-wall imaging is used by building construction workers to search for hidden elements such as pipes, wiring and studs in a wall. Security perimeters use the radar properties of UWB to establish a virtual 'fence'. Intruders crossing the fence are detected and an appropriate alarm is issued. Of these general imaging applications, GPR is referenced most often in the regulatory processes and other literature. The volume of GPR units is expected to be very much lower than either HDR or LDR.

1.1.4 Automotive radar

The fourth and final major category of applications is automotive radar for collision avoidance. Figure 1.2 is a graphical representation of the various collision radars that are planned for cars. In these applications, UWB is used by the car as a radar to trigger automated braking when a collision is judged to be imminent. By forcing the car to brake involuntarily, the force of the impact is reduced substantially. Automotive manufacturers expect that this new capability will save a significant

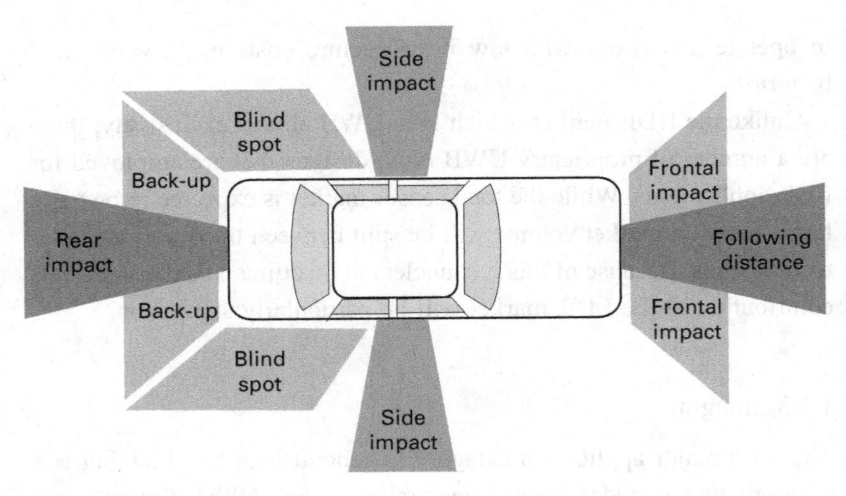

Figure 1.2 Automotive radar

number of lives. Like ground-penetrating radar, automotive radar is developed for its imaging properties and is not used for communications. Additionally, automotive radar is the only form of UWB that presently uses the 24 GHz band. It does not use the 3.1–10.6 GHz range that is employed by HDR, LDR and GPR applications.

1.2 Next-generation HDR applications

The HDR applications described in the prior section are based upon a very simple view of the world. Conceptually, devices were connected by a wire. First-generation applications are based upon a logic wherein the wire is removed and is replaced with a radio in order to improve the aesthetics of an installation and to prevent customers from having to keep up with the wires for their mobile devices. In the second generation applications, connectivity is all about connecting mobile devices while making the process somewhere between simple and invisible from the user's perspective.

To begin this topic, consider a trend that has been evident in computers for the last 50 years or so. Computational platforms are becoming smaller, cheaper and more capable with every generation. So far, the industry has moved through the mainframe, mini-computer, desktop and laptop generations. The next generation, which is emerging now, is the handheld computer–phone. It will be a converged product that includes the intelligence of a computer, communications capability of a phone and the entertainment capabilities of a music player, video player and gaming device all rolled into one.

Each generational step brings new capabilities to the consumer that were not possible before. In the handheld generation, the new capability is full mobility. Consumers will soon be able to carry huge volumes of data, programming and entertainment content with them wherever they go.

Look at the same picture from another perspective. The existing generations of PCs, set-top boxes and game consoles have been bound together by the Internet into a very large communications web. As the handheld computer platform emerges, it needs to be bound into that network as well to perform the functions that it wishes to execute. The cellular community imagines that this activity will take place across the wide area network (WAN). Unfortunately, the performance limits of that network make the WAN insufficient to handle the full job. For instance, transferring a large volume of high-resolution photographs, a full-length video or a state-of-the-art game program would take an extremely long time across the cellular network. Additional connectivity is required to do the job. The PAN (personal area network) is the principal path through which much of this work gets done.

To understand in greater detail about the applications that will emerge from connecting the mobile generation, one needs to get a feel for what consumers wish to do with the device. To begin, consumers wish to be able to work cooperatively with stationary devices such as the STB, media centre and the PC to archive data that is not being used immediately. Most individuals have far more photographs than they carry around in their wallets. Likewise, people will archive most photographs on a stationary device and only carry the most meaningful photos with them. In the case of content, only that material which the user is inclined to consume in the near future will be carried on the mobile platform. The remainder will be archived on stationary devices. The present wire-replacement strategy would seem to work reasonably well in this case.

Looking closer at the mobile device, it doesn't take long to realize that the user interface is pretty meagre. The keypad is inconveniently small, as is the display. Asking the consumer to enter file names and directory addresses will quickly become intolerable. To compensate for this, there are several things that can be done. First, if the mobile device is in range of a more capable stationary device, like a high-definition display or a computer, it will be desirable for the mobile device to be able to use them. In the case of the display, imagine the consumer touching the display on a phone (using near-field communications to establish the link) and then displaying photographs on the large screen. Or, imagine the same consumer sitting down at an Internet café. His hard drive is linked to the stationary PC. From his perspective, he has effectively turned the rented PC into his own system.

Both of these applications require the high-speed link and short range for which a PAN is optimized. But there is more. In many instances, there simply isn't an available keyboard to use to make transactions. Other methods to minimize the amount of interaction required between the user and the keyboard are necessary.

Imagine if a consumer were to walk up to a kiosk or an STB and wishes to download a video onto a mobile device to watch on the train into work. The movie is selected manually, but instead of entering information about credit-card numbers or file-storage locations, the transaction simply occurs. This would be an example of using intelligence tightly coupled with the communication links to make use of the handheld device convenient. While this type of capability could be implemented at an application level, it would also be possible to implement it as a standardized set of primitives that operates at a lower level.

With this type of approach a number of new applications become convenient. Imagine a mobile device subscribing to a music service. The mobile device is placed next to a PC each morning and its music is updated. The device is then carried out into the car where the music is then copied onto a car radio. If a song is skipped or repeated, these actions are recorded and transferred back to the mobile, the PC and ultimately to the music service, where a preference profile is developed for the user. Tomorrow's music selection will take these preferences into consideration.

If one were to look at these actions from a high elevation, what is occurring is the integration of this large and growing cloud of mobile devices into the network web that exists. Physically connecting them is only the first part of the problem, which is resolved by wire replacement using a PAN. It is also necessary to solve the deficiencies of the mobile platform simultaneously to get consumers to actually want to use the connected capability that is enabled. While it is indeed possible to require every application to resolve the problem for itself, it is far more efficient to establish standardized routines that can be called to do the job. This later approach also has the benefit of creating a more uniform expectation from the consumer of how to interact with the device to get a task done.

1.3 A brief history of ultra-wideband

This text is going to spend the overwhelming percentage of its time investigating the HDR portion of the UWB market in detail. This is expected to be the portion of the market with the greatest number of UWB units sold and the largest number of manufacturers developing products. If this is insufficient reason to concentrate our efforts there, both authors show a certain bias toward this sector (owing to their active involvement therein) and have elected to admit that fact openly.

The remainder of this chapter will be dedicated to providing a very cursory history and developmental background of UWB. A more exhaustive history, including amusing anecdotes about the exploits of the authors, was deemed not to be within the boundaries of *The Essentials of UWB* and has therefore been reduced to the following.

Ultra-wideband started life in the late 1960s as a military technology radar which was capable of penetrating leaves and other foliage. It was also used in military applications for covert communications. In this application, the military took advantage of the fact that UWB signals were spread across a very wide bandwidth and could be made to appear as noise to most interception equipment. Because of this military focus, the foundational work on UWB was done for the first few decades by companies involved in military provisioning. By the 1990s, UWB had completed its stint in the military and was ready to begin the transition into the commercial world. Unfortunately, there were certain regulatory and business obstacles, which prevented UWB from being immediately deployable. More simply put, it was not legal anywhere on the planet and could not be built for less than a few thousand dollars per device. At this same time, silicon-device capabilities made it possible to build low-power UWB devices very inexpensively. This resolved the cost problem that prevented it from becoming commercial. The only thing standing between UWB and a life in the private sector was the lack of regulations.

By the time that UWB came to the attention of the FCC in 1998, the worldwide regulatory structure was approaching 80 years old and was founded on the concept that spectrum should be allocated in small frequency chunks, which are geography specific and possess only the most simplistic concept of spectrum sharing. As with most 80-year old technologies, the system was beginning to show its age. It was not designed to incorporate contemporary advances, such as intelligent use of spectrum (through radios that could employ intelligence to avoid interference), and the need for extremely high-data-rate communications. The old regime was also not designed to contemplate the use of very wideband transmissions, as this would tend to consume the accessible spectrum too quickly. At the inception of the regulatory processes, high-data-rate communications meant a quick Morse-code operator tapping away at speeds near 100 words per minute. Interference avoidance meant using a different band on which to transmit.

When the public discussion about UWB began, a UWB signal was envisioned to be based upon a series of pulses. Each pulse would appear as a very sharp spike when viewed in the time domain. This would, in turn, create a very wide signal which propagated across a large number of frequencies when viewed in the frequency domain. The sharp edges on this pulsed signal made it possible to gauge distances accurately between the source transmitter and any reflection target. Figure 1.3 attempts to show this effect graphically. A fast rise time creates a shorter period of uncertainty about when the pulse was

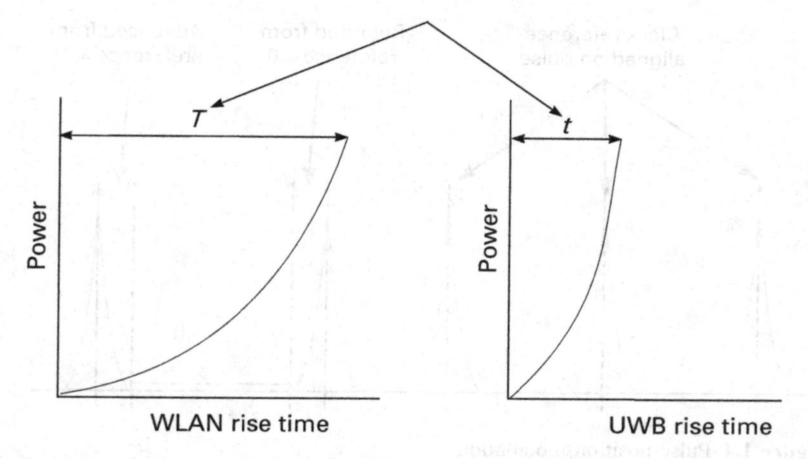

Figure 1.3 Correlation between fast rise time and location accuracy: a fast rise (spike) reaches full power more rapidly, reducing the uncertainty of when the spike was initiated. Distance from the emitter to the receiver is calculated as $D = \text{time} \times$ speed of light. Reducing the uncertainty of time reduces the uncertainty of distance.

transmitted and, therefore, more accurate location resolution. The ability to resolve location is what enabled UWB to function as a radar and also made it possible to establish the position of UWB transmitters for location-tagging purposes.

The first commercial effort at modulating the signal to provide communications with UWB was based upon the shifting of the pulse in time. Pulse-position modulation (PPM) was based upon a known time reference point. Figure 1.4 shows how PPM functions. If the pulse were transmitted before the reference, it would represent a digital zero and if it were transmitted after the reference point, it was a one. It was a beginning, although it would ultimately prove not to be a particularly good start.

As the FCC hearings got under way, the entrepreneurial spirit seized several individuals who proceeded to found start-up companies. This was immediately followed by high-spirited and aggressive spending of venture-capital funds on a large number of promotional activities designed to make UWB legal and to raise additional venture-capital funding. The corporate evangelists hired to bring about the legality of UWB advocated this new technology to the US Congress and the

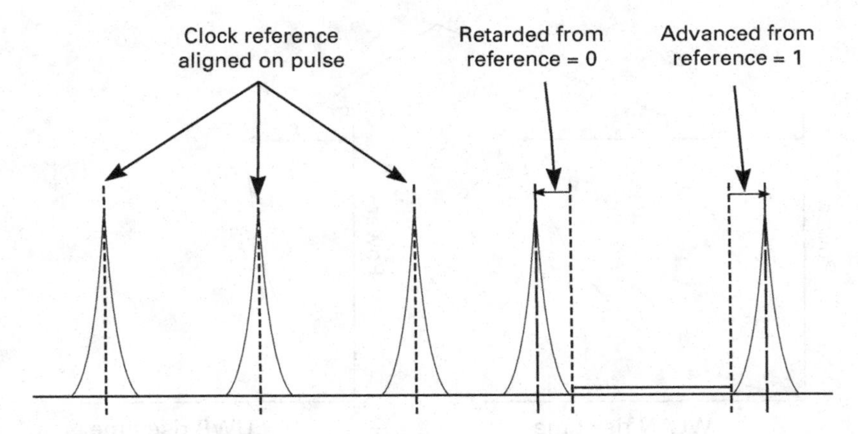

Figure 1.4 Pulse-position modulation

Federal Communication Commission (FCC). It was falsely advertized as completely non-interfering to existing radio systems and therefore could be used as 'free' spectrum that was potentially outside the normal rules of physics as they are commonly known.

In fact, this was not the case. Ultra-wideband emitted RF energy just like any other radio, but simply across a wider spectrum. However, when the FCC investigated these claims, they determined that UWB was not the cure-all that was being advertized, but it might be used as a mechanism to improve some of the shortcomings of the existing spectrum management policy. Ultra-wideband could be made to be relatively non-interfering by using very low power. The UWB allocation would make up for some of the shortcomings of lower power by providing exceedingly wide bandwidth. The logic of this approach is that it enabled the creation of a very short-range radio that had the very high throughput potential that was needed by industry. With this concept, the FCC began regulatory proceedings to create a UWB allocation.

Simultaneous with the FCC's work to create an allocation, the industry went to work on improved modulation techniques to try to produce a system that had lower interference potential and better performance characteristics than the original PPM method. During early development of PPM, it became clear that the radios had difficulty in adequately resolving the time shifting necessary to decode the data.

In-building environments have a large number of reflections, or multipaths, which were challenging to separate from the main signal path using PPM. Early regulatory investigations that occurred as part of the FCC proceedings also demonstrated that PPM had a higher-than-anticipated potential for interference. In response to these issues, the industry developed, evaluated and refined several approaches that outperformed PPM. The majority of this development work was done as part of the IEEE 802.15.3a committee proceedings.

One approach, which was proposed to the IEEE, was called 'direct sequence'. This technique treated the available spectrum as one (eventually two) large blocks. Carrier-division multiple access (CDMA) techniques were then borrowed from the cellphone industry to allow for the spectrum to be divided into logical channels. This approach to the use of the spectrum had an assumption that the spectrum allocated was contiguous and had uniform rules applied throughout the band. As the regulatory environment began to become more visible over time, it became clear that this approach would be expensive and difficult to implement.

Eventually, the industry 'settled' upon the second major proposal which was developed during the IEEE work and was called the multi-banded orthogonal frequency-division multiplex (MB-OFDM) technique, which is described in detail in Chapter 3. These OFDM-based technologies are able to combine the energy across multiple signal paths in a highly reflective building environment to create a more robust signal than competing technologies. The multi-band portion of the design divided up the available spectrum into channels, which could operate and be controlled independently of each other. It was believed that this feature would make the radio design more flexible in the face of varying regulatory requirements around the world. By contrast, a direct-sequence design uses a radio architecture that divides the spectrum into two very large bands and that has a multi-element receiver (referred to as a rake receiver) which dedicates a rake finger to each of the major signal paths to combine energy and likewise outperform PPM.

To say that the industry 'settled' upon MB-OFDM is a rather polite term to describe an outcome while simultaneously overlooking approximately three years of intense standards controversy that went on between the MB-OFDM and direct-sequence supporters to get to that result. This period of conflict made for excellent fodder for the industry press and provided a testing ground for UWB-radio design that forced a great deal of improvement in the final MB-OFDM radio being deployed today. Obviously, this selection process has been somewhat shortened here.

While the standards processes were working toward becoming 'settled', a similarly contentious discussion was taking place in the FCC proceedings now in full swing. In 2000, the FCC issued its Notice of Proposed Rulemaking (NPRM) [3] on UWB. This promptly opened the floodgates to comments from industry both pro and con, which went on for the next two years.

Organizations that already had spectrum were somewhat less than enthusiastic about the FCC's proposed changes than the UWB proponents whom the FCC intended to enable. In most cases, these groups viewed the existing spectrum allocation as 'their spectrum' whether they had won the license in an auction or were given it by the FCC, and were willing to oppose actively any change to the spectrum use, which they viewed as having the potential to degrade the value of their license. This opposition came out during FCC proceedings on the topic and is expressed in more than 700 written responses from industry. [4]

In retrospect, it is clear that the UWB proceeding was inevitably going to be one of the more hotly contested regulatory decisions. The FCC's proposal allocated an underlay across more than 7 GHz of spectrum. This immediately polarized a large percentage of the existing spectrum holders as each group felt compelled to defend its existing allocation. Additionally, the UWB decision advocated for a revision to the logic that was historically employed in spectrum management. Instead of a fairly hard and deterministic allocation of spectrum into fixed frequency blocks, as was the historical norm, the ruling suggested a trend toward a more shared use of spectrum. It was a trend that incumbents felt inclined to resist.

Despite spirited opposition, on 14 February 2002, [5] the FCC gave the UWB industry a Valentine's Day present in the form of a favourable ruling authorizing the use of UWB. A second Report and Order in 2004 [6] further refined the position. While these two rulings set policy in the USA, it only acted as the starting gun for the debates in the International Telecommunications Union, as well as regulatory and standards bodies in Europe, Asia and the rest of the world.

At the time of this writing in 2007, the industry is finally moving to open the market that the FCC originally envisioned so long ago. The initial regulatory proceedings around the world are slowly drawing to a close. (Further refinement will no doubt follow for several years yet.) By the end of 2007, it is expected that regulatory approval will be granted in the USA, Europe, Japan, Korea, China, Canada and possibly one or two other countries. The first UWB products from more than a dozen manufacturers are expected to be certified and shipping in approximately the same timeframe.

1.4 Summary

Before leaving the introduction, a quick description of the text's organization is in order. This chapter and Chapter 2 describe the applications for which UWB is being applied as well as a discussion about why UWB is uniquely qualified to be deployed against those applications. Chapters 3 and 4 are intended to provide the technical foundation by describing both the radio itself as well as the protocols which are added on top of the radio by several different industry organizations. Chapters 5, 6 and 7 are intended to provide data about the organization of the industry in terms of standardization relationships and special-interest groups which will be used to certify and promote UWB. Chapter 8 is a description of UWB regulations. This chapter is intended to shed a little light on the processes and positions of groups participating in the regulatory proceedings. Chapter 9 discusses the market trends and other business aspects of UWB. It sets expectations for issues such as market consolidation, single-chip integration and other likely developments. Chapter 10 is a discussion of the regulatory process that led to the legalization of UWB. It describes much of the regulatory philosophy and logic that were critical in the development of regulations. Chapter 11 is entitled 'The tragedy of the commons'. This name is applied in a regulatory context to describe mismanagement of spectrum that is publicly held. For UWB not to become a victim of mismanagement, it is necessary to manage the bandwidth-consumptive applications, which are both its bread and butter as well as its greatest risk. Hopefully, this will aid industry participants in understanding a necessarily complex and somewhat opaque process.

References

[1] Siwiak, K., *Ultra-wideband Radio Technology*, Wiley, 2004.

[2] Aiello, R., and Batra, A., *Ultra Wideband Systems Technologies and Applications*, Elsevier, 2006.

[3] www.fcc.gov/Bureaus/Engineering_Technology/Notices/2000/fcc00163.doc.

[4] *Testimony of Julius P. Knapp before the US House of Representatives Committee on Energy and Commerce Subcommittee on Telecommunications and the Internet*, June 5, 2002, p. 5, www.fcc.gov/Speeches/misc/statements/knapp060502.pdf.

[5] www.fcc.gov/Bureaus/Engineering_Technology/News_Releases/2002/nret0203.html, accessed 4 October 2007.

[6] http://fjallfoss.fcc.gov/edocs_public/attachmatch/FCC-04-285A1.pdf, accessed 4 October 2007.

2 Matching UWB to HDR applications

Any successful new technology can be described as a combination of features that allow the technology to perform a given application better than those technologies that precede it or that enable new applications to be performed. The consumer has the final word in the success of a technology. If the product that manufacturers are trying to sell to the consumer does not convey a strong sense of benefit or sex appeal, the product becomes a wallflower on the back of store shelves.

In this chapter, the discussion will centre on the features of UWB, a comparison of these features with competing technologies and the emerging applications that demand the improved performance that UWB provides. With UWB, the principal features of interest include speed, cost, location resolution and power consumption. Each of these characteristics will be covered separately.

2.1 Speed – specifying UWB

The exciting new applications that are emerging now or will emerge over the next few years will demand, more than any other single attribute, extremely fast speed. Speed is required for one of two reasons. Either the application involves a large file transfer, such as the download of a Blueray DVD (50 GB), [1] or high-resolution video streaming (Displayport up to 11 Gbps). [2] In the case of large file transfers, speed translates into consumer wait time. Instead of taking hours to transfer the contents of a DVD on to an 802.11a wireless local area network (WLAN), a UWB link could be expected to reduce this time to minutes in the current generation, moving to two minutes in the second.

Here is a quick calculation to show this result. In the case of 802.11g, the calculation would proceed as follows:

Step 1: 50 000 000 000 (billion or Giga) × 8 (bits in a byte) = 400 000 000 000 bits to transfer.
Step 2: 54 Mb/s (signaling rate for 802.11g) × 0.4 (system overhead) = 21.6 Mb/s (peak transfer rate).
Step 3: 400 000 000 000 bits/21 600 000 b/s = 18 518.5 s/60 (s per min) = 308.6 min/60 (min per hour) = 5.1 hours to transfer, assuming peak rate.

The same equations, run for a UWB device would proceed as follows:

Step 1: 50 000 000 000 (billion or Giga) × 8 (bits in a byte) = 400 000 000 000 bits to transfer.
Step 2: 480 Mb/s (signaling rate for 1st generation UWB) × 0.85 (system overhead) = 408 Mb/s (peak transfer rate)
Step 3: 400 000 000 000 bits/408 000 000 b/s = 980.4 s/60 (s per min) = 16.3 min.

Until very recently, the concept of large file transfer applications has largely been a PC phenomenon. Files are transferred to hard drives, printers and scanners. The PC had a very large storage relative to other devices and access to high-performance communications links, which made these functions viable.

This capability is moving beyond data files and the PC to other areas. Most of this movement has been enabled by the introduction of very large capacity, but also by very small disk drives and very high-density non-volatile memory. Digital cameras have very quickly evolved to multi-megapixel images. Because the cost of photo processing has been effectively removed, there is little reason not to take many more pictures and simply throw away anything that doesn't work out. This results in Gigabits of personal photos that must be stored and moved.

The emergence of music players and personal video players is another example of this trend. One product on the market today has an 80 GB capacity capable of carrying thousands of songs and full-length videos as well. While the limited screen size prevents users from getting a full cinematic experience, it is more than acceptable for viewing on trains, planes and during downtime that would otherwise be spent waiting.

This trend toward greater personal storage is far from satisfied. The first Terabyte hard drive in a portable form factor was released in 2007. Holographic and three-dimensional storage techniques that are in prototype and research stages today make it probable that the exponential growth in this field will continue for a number of years yet. Video quality in terms of higher resolution, larger display sizes and color depth also make it likely that the content being developed will make increasingly higher demands on storage. But clearly, increased file sizes and storage capacities are only functional if the content can be moved onto and off of the platform with relatively little effort.

Streaming video is the second major class of application demanding very high transfer rates. Streaming video is defined as video content being delivered to the display with minimal buffering at a rate wherein the display matches human consumption rates. Streaming video principally occurs in television displays and on computer monitors.

Uncompressed video is an extremely demanding application in terms of the amount of bandwidth required to achieve the image quality and visual consistency to which consumers have become accustomed. Just to provide a few samples, a 1080 p (high definition) video stream requires approximately 1.5 Gb/s if the data are uncompressed. If the video stream is compressed in an MPEG 2 format, the bandwidth required will vary with the video content, but will range from approximately 19–24 Mb/s, while MPEG 4 compression will reduce that bandwidth to approximately 10 Mb/s. This is the current state of the art in broadcast video.

If one were to look at computer-monitor technologies, the bandwidth required is significantly higher. The latest technologies (high-definition multimedia interface (HDMI) and Displayport) have upper bounds of about 11 Gb/s. Unlike broadcast video, monitor streams are not easy to compress. The compression processes used in MPEG techniques requires very-high-power computational resources and significant amounts of time to process a video program into a compressed format.

Until it is possible to perform similar compression in real time, the ability to reduce bandwidth requirements in monitor applications will remain limited.

Additionally, the amount and cost of the processing required to decompress these streams is not insignificant. Monitor and display manufacturers, who have strong price pressures, have an incentive to use uncompressed video streams wherever it becomes technically feasible to do so.

This just describes the situation as it stands today. Consider where video is going over time. Consumer electronics manufacturers are attempting to create as realistic an experience as possible for viewers. They want consumers to feel like they are standing on the summit of Everest while watching the sunset in all its brilliant glory without any of the expense, effort or frostbite. To achieve these objectives, they need to build larger video systems with higher resolution, frame rates and colour depth.

To date, the evolution of video has been constrained by the size of storage devices and transmission pathways. These limitations are changing with the introduction of fibre-optic television networks and blue-laser DVD storage. Because of these changes, it is reasonable to expect that streaming video content will continue to become more bandwidth intensive for the foreseeable future.

At this time, UWB and almost any wireless technology are inadequate to handle uncompressed video of the calibre available today, much less what is on the drawing boards. But, with work on compression techniques and other compensating methods, it is possible to use UWB to address some of the lower-resolution video requirements. (There are a number of regulatory caveats to this statement, discussed in Chapter 8).

High-data-rate UWB devices entered the market in the first half of 2007 [3] with an advertized data rate of 480 Mb/s (at 3 m) and 110 Mb/s (at 10 m) with plans to move the maximum to greater than 1 Gb in the second generation. Ultra-wideband compares quite well with the specified rates of 802.11b (11 Mb/s), 802.11a/g (54 Mb/s) and Bluetooth 1.1 (700 kb/s), but appears somewhat anaemic when compared with the

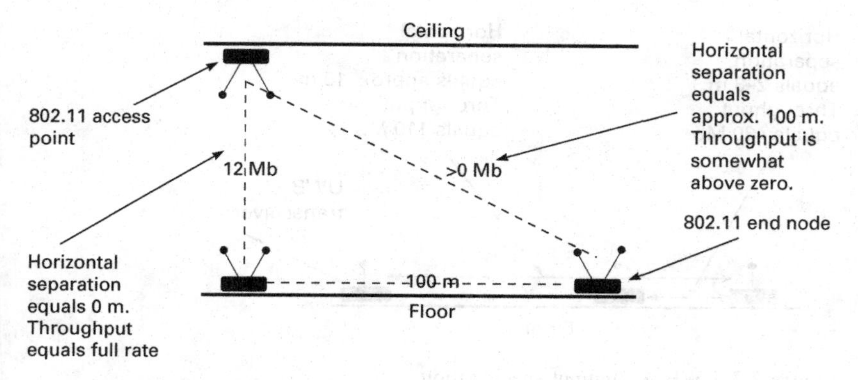

Figure 2.1 802.11b throughput specification

specified signaling rates of 802.11n (1100 Mb/s). However, things are not always as they seem.

Even though these numbers appear to be directly comparable and will be used that way in the press and by consumers, they are not. The numbers for 802.11 (WiFi) technologies bear the effects of a long history of specmanship. Here are a few of the finer points on the subject. The 802.11 local area network (LAN) standards specify the maximum signalling rate and the maximum distance but fail to mention that these goals cannot be achieved at the same time. Stated another way, 802.11g is rated to signal at 54 Mb if a range of effectively zero metres is obtained by having the user stand directly beneath an access point with nothing blocking the path between the two. Figure 2.1 demonstrates this. Standard 802.11g is also rated to transmit at a range of about 30 m, but this range is only possible if the transmission occurs in a flat open field devoid of obstructing obstacles and vegetation and with a data rate that is more commonly associated with communication by semaphore flags. Specifying the extreme points of performance has historically been done to enable manufacturers to position 802.11g to consumers with the most positive specifications possible. In practice, customers usually see between 20–40% of the specified values. (WiFi throughput is 20–40% of rated value.) So an 802.11 link specified at 54 Mb/s will be most likely to provide 11–22 Mb/s under favourable conditions. An 802.11n system fully outfitted for speed will actually

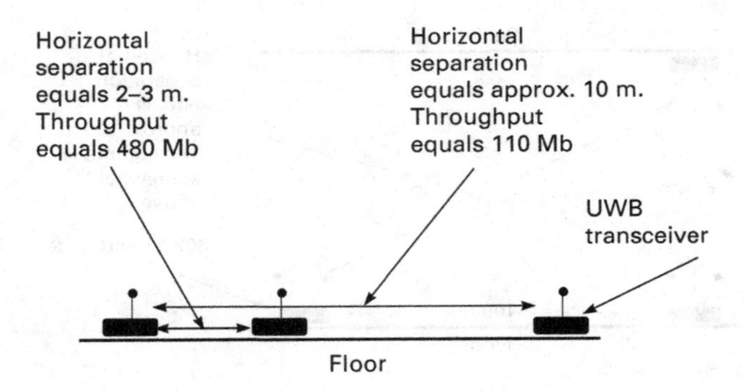

Figure 2.2 UWB throughput specification

deliver 220–440 Mb/s to the user under favourable conditions while positioned beneath the access point. This rate will decline as the user moves away from the access point.

By contrast, UWB is specified for data rates at specific distances (480 Mb/s at 3 m and 110 Mb/s at 10 m), which are intended to represent probable-use cases. This provides a description of the degradation due to distance that the user will experience through separation distance. Figure 2.2 demonstrates that UWB specifies its throughput as a function of the range.

Additionally, the overhead incurred by the media-access control (MAC) in UWB is much less than in 802.11. The MAC overhead introduces approximately 15% overhead for video and approximately 40% for asynchronous traffic. This compares well with 802.11's 60–80% overhead rate. One must also be aware that 802.11n's peak rate of 1100 Mb/s will only be possible for a high-end multiple in, multiple out (MIMO) implementation using unusually wide (40 MHz) channels, which has significant trade-offs in terms of size, cost, power and the number of simultaneous users supported. In addition to this maximum throughput configuration of 802.11n, there are also flavours that use 20 MHz channels or different antenna configurations. These versions trade lower throughput in exchange for reduced power consumption.

One reason that different approaches to specifying throughput rates have occurred is the traffic that was expected to be carried. In the case of 802.11, it was originally anticipated that the data crossing the LAN would not be time critical. Traffic was principally made up of email, text reports and photographs. The network was designed to be an Ethernet replacement and, like Ethernet, it used a philosophy of best-effort delivery of the data. If there were dropped packets that had to be repeated, owing to a marginal communication link, so be it. Under this logic, specifying the maximum performance points allowed the 802.11 standard to compare favourably with wired Ethernet. By contrast, UWB was designed with the concept of time-critical services such as video and audio in mind. This logic required the implementer to have a more realistic expectation of the actual performance of the radio under likely use scenarios. And while the UWB method of specifying throughput does not get the problem completely solved either (because of other variable overheads), it is a step in the right direction. A full analysis of the actual performance of UWB or any other wireless technology would need to take into account a host of inefficiencies, such as media-access-control overhead, protocol overhead, attenuation losses, etc. Some degree of simplification of the numbers presented is almost unavoidable.

A broader comparison between UWB and 802.11 is beyond the scope of this document. For now, recognize that a direct comparison between the two technologies can be extremely involved and depends heavily upon the intended application that the user intends to employ. The one thing to take away here is to beware of the unqualified claim.

One factual assertion that can be made regarding the throughput of UWB is that it is reasonable to expect it to increase significantly over time relative to its starting point. The first version of UWB is advertized at a maximum rate of 480 Mb/s today. A version that is anticipated to be at or above 1 Gb/s is already part of the standard development plans. Several companies have even demonstrated this capability in the form of prototypes or proprietary modes. Ultra-wideband has yet to explore the known options for improvement, such as higher-order modulation, antenna diversity or intelligent antenna structures to improve its performance. When this is done, it is reasonable to expect UWB throughput to grow substantially, with a realistic peak level expected to be between 2–5 Gb/s.

2.2 Low cost

The second principal asset that UWB has to its credit is very low cost. Several different factors make UWB a very inexpensive radio. First is the very low emitted power level authorized for UWB. Ultra-wideband emits less power intentionally than a stereo, drill or television is allowed to emit as unintended noise. These very low emitted-power levels mean that it is not necessary to use an expensive external power amplifier as part of the radio design.

Additionally, the design of the UWB radio is largely digital. This means that the overwhelming portion of the radio scales down as the size of digital circuitry declines over time. Analogue devices, which make up a larger percentage of other radios, do not scale as well with new silicon processes. These components must essentially be redesigned with new process generations.

Because digital components scale readily, smaller pieces of silicon are required to build a radio for each new silicon process step and the costs of the device decline proportionately to the silicon area consumed. Process improvements are normally driven by high-performance and high-volume parts, such as microprocessors and memory chips. Ultra-wideband radios piggyback on these improvements with the minimal additional effort invested.

Economies of scale and competition will also help erode UWB prices. Ultra-wideband entered the market at the middle of 2007 at a price point of about $10–15 for the chipset when delivered to an integrator in volume. Over the course of the next two or three years, the price of UWB can realistically be expected to fall below $5. This follows a cost decay curve that has been historically true for Bluetooth chips as well as 802.11 LAN chips. Cost improvements are driven through integration and process improvements. They begin by employing multiple chips to build the radio as a method of simplifying the radio design. These chips are then merged together to reduce the costs. And finally, the single merged chip goes through process improvements that reduce the amount of silicon required. This path has been historically valid in both Bluetooth and Wireless LAN chips.

It is reasonable to expect UWB development to follow a similar trajectory.

Once a very high speed, low-cost UWB radio is available, it will be possible to address the wire replacement applications in the PC and mobile handset markets. Printers, scanners, external hard disks and cameras can employ UWB to eliminate the cable clutter in homes and offices. In the PC markets today, peripherals are connected to the PC through a variety of wired connectors. One of the major wired buses is the universal serial bus (USB). Ultra-wideband is being used by the USB Implementers Forum (USBIF) to extend the USB franchise into the wireless space with Certified Wireless USB (CWUSB or WUSB). This essentially means USB protocols which have been optimized for a wireless environment and then run on an underlying UWB radio.

In the mobile-handset market, the Bluetooth Special Interest Group (SIG) is doing something similar to Wireless USB. It is overlaying Bluetooth protocols on a base UWB radio to create a next-generation Bluetooth radio. In this case, the original 2.4 GHz Bluetooth radio replaced the wire between an earpiece and the mobile phone and so is reasonably classified as wire replacement. Ultra-wideband increases the throughput possible with Bluetooth and thereby expands its usefulness. In both of these cases, the low prices possible with UWB are expected to make it economically viable for UWB to be included on mobile handsets and personal computers.

As UWB reduces in price, it is able to be built into less expensive devices. As an example, a $15 UWB radio placed into a $1000 PC is not insignificant, but it is manageable. If, one were to try to justify putting the same radio into a $50 mobile phone, it is unlikely that this could be done. Clearly, if the percentage cost of the radio compared with the device containing it is too high, it becomes difficult or impossible to justify the addition.

Conversely, as the price of the radio falls, the number of devices which can afford to include a UWB radio expands. If one considers the ongoing trend towards small, mobile, smart devices, it becomes clear that reducing the price of the radio becomes an absolute necessity to participate in the emerging markets.

2.3 Location

The third major feature of UWB that enables interesting new applications is the ability for high-resolution location of a UWB transmitter. Stated simply, with the right network in place, it is possible to identify the location of a UWB transmitter to a resolution of a few centimetres. This enables a variety of applications that depend upon location, such as tagging goods in a warehouse, asset control and monitoring in a retail store, and access control in an office environment.

Employed in a tagging application, location can be used to resolve the location of goods within a three-dimensional area such as a warehouse or factory. This is done using radio tags that emit periodically. A network of receivers is then set up in such a manner that they are able to triangulate and resolve the location of the originating transmitter. The network has to be set up explicitly for the purpose of triangulation for this to be efficient. Figure 2.3 shows a very simple location network. In it, each circle represents the reception range of a receiver. When a transmitter can be simultaneously heard by three (or more) receivers, its location can be resolved unambiguously. If one wished to resolve height as well, a fourth receiver would be required.

By way of example, consider a large, automated warehousing operation. Each pallet in the facility contains a UWB tag, which transmits its serial number. Some applications may require that the tag have a receiver as well, but frequently this is sacrificed to maintain longer battery life. A grid of receivers is positioned around the warehouse so that three or more receivers can see any transmission. When an automated fork-lift moves the pallet, the system is able to get a physical confirmation that it ended up where it was supposed to go. This technique can also be used very quickly to set up military warehouses in theatres of operation with very minimal infrastructure.

A second interesting example involves a retail store. Retailers have found that customers are far more likely to purchase goods if they can handle them. Placing expensive goods under lock and key actually reduces sales volume. To reduce this problem, it is possible to employ

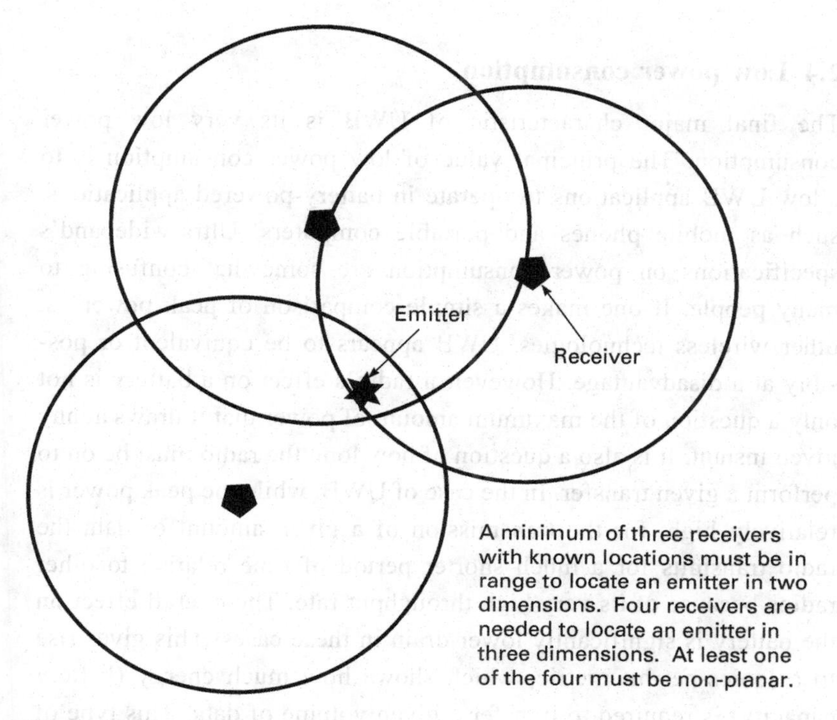

Figure 2.3 Location triangulation

location tags. Imagine a retail clothing store that has a women's coat department. All coats are supposed to be purchased at a register within the department. Each coat has a location tag attached. As long as the coat remains within the perimeter of the department, nothing happens. If a tag moves beyond those boundaries, a security camera is focused on the tag's location and security personnel are both alerted and shown a visual display of the activity. As you can see, installations using location frequently have rather particular requirements that do not lend themselves well to mass-market solutions.

Most SIGs addressing HDR applications, such as WUSB or Bluetooth, do not choose to use the location capabilities of UWB. As WUSB and Bluetooth devices usually involve communication between two individual units for the purpose of wire replacement, location is less critical to the application and so is not employed.

2.4 Low power consumption

The final major characteristic of UWB is its very low power consumption. The principal value of low power consumption is to allow UWB applications to operate in battery-powered applications, such as mobile phones and portable computers. Ultra-wideband's specifications on power consumption are somewhat confusing to many people. If one makes a simple comparison of peak power vs. other wireless technologies, UWB appears to be equivalent or possibly at a disadvantage. However, a radio's effect on a battery is not only a question of the maximum amount of power that it draws at any given instant. It is also a question of how long the radio must be on to perform a given transfer. In the case of UWB, while the peak power is relatively high, for the transmission of a given amount of data the radio transmits for a much shorter period of time relative to other radios because of its very high throughput rate. The overall effect on the battery is significantly lower drain in these cases. This gives rise to a joules-per-bit metric, which shows how much energy (battery capacity) is required to transfer a given volume of data. This type of metric is much more appropriate for understanding the effect of the transfer on battery life. When measured in this way, UWB performs quite well relative to potentially competing technologies.

The power-per-bit argument in favour of UWB falls short if the application is continuous streaming instead of file transfer. A UWB radio in a streaming application will remain on and consume greater power than other radios unless the designers have implemented the radio in such a way as to allow it to fall into a sleep state frequently. As an example, consider the streaming transmission of an MPEG 2 stream running at 20 Mb/s on average. The UWB radio is operating at 480 Mb. There is sufficient time for the radio to power-down receiver components between packets if it has been designed to operate in this manner. In this situation, the extra bandwidth of UWB allows it to send or receive a packet quickly and then sleep for a period until the next packet.

2.5 Personal area network architecture

With this understanding of UWB's strengths, the next step is to establish a common understanding of the network architectures in which UWB will be used. In this regard, one point cannot be made too strongly. Ultra-wideband is a superb personal area network (PAN). It is only a very marginal local area network (LAN).

2.5.1 Range does not equal goodness

There is a belief among many radio designers that range = goodness. They have an instinctive belief that they should be attempting to cover the largest area possible. If possible, they attempt to do this with a single transmitter. In other cases, they attempt to cover a large area by connecting a mesh of short-range devices. This approach allows for broad sharing of expensive resources, such as high-speed printers. It reduces capital expenditure, but requires that employees are trained in how to address communication to the device using potentially obscure names and codes. This is reasonably sound for business and industrial environments, but is sub-optimal for more intimate spaces like the home. In the home, the structure of communications tends to be more concentrated. Single-level or 'one-hop' networks can employ much more simplistic communication methods that are more intuitive for untrained consumers.

As a side benefit, one-hop networks take advantage of another useful trend. In a fixed area and with a fixed bandwidth, the total capacity of a network can be increased by creating a number of small, independent communication cells as opposed to one large cell or a mesh of small cells where a transmission must be repeated. Based on this principle, the best method for allowing the greatest traffic in the home is to create a number of small independent cells that reuse the same spectrum in different parts of the house. Communication between cells is conducted across a wired backbone.

The instinctive urge of developers to maximize the area covered causes them to experiment with inappropriate uses of UWB. For instance, it is technically possible to employ UWB with directional antennae to go through one or two walls. This maximizes range but does so at the cost of individual and aggregate throughput as well as ease of use. Likewise, other developers will attempt to organize UWB devices into mesh networks to cover a larger area. This structure is inefficient because it requires the data to be repeated at each hop in the network. Instead of transmitting data once, it may be transmitted two or three times. The total capacity of the network is traded off for coverage or range. For the broader HDR markets, these experimental attempts are simply a distraction. They use UWB in applications for which other radio designs provide superior performance or design flexibility.

2.5.2 The natural stratification of wireless networks

If one looks at the available technologies in the market, there is a natural stratification that occurs. Wide area network (WAN) or cellular technologies, such as Global System for Mobile Communication (GSM) or code-division multiple access (CDMA), are optimized to cover very large areas with low to moderate data rates (approx 2 Mb or less per user) with devices that may move through the network at the speed of an automobile. Local area networks are designed to support substantially higher data rates (>54 Mb) over areas of less than 100 m while being required to penetrate several walls in a home or office. Personal area networks are designed to operate at 10 m or less in a single room with peak data rates that are potentially several hundred megabits per second to several gigabits. Each of these radios is optimized for the environment in which it exists. While it is possible to push a LAN radio into a WAN context as much as it is possible to push a PAN radio into a LAN context, these are not optimum designs. The network does not perform as well and customer applications are not serviced as effectively as when the application is matched to an optimum radio.

The 10 m PAN radio architecture has certain advantages, which are difficult to replicate when the scale (range and network complexity) is increased. First, the number of devices that one might communicate with in a single room is limited relative to the number of devices that one might encounter in an enterprise LAN network. This fact can be used by developers to create designs that are easier for the consumer to use. If one wished to get a camera to talk to a printer, this might be as simple as touching the two together. By contrast, if the network size has the potential to cover the floor of an office building, the association models cannot assume that the two devices are in the same room, where touching makes sense. Instead, it is necessary to use some form of addressing that may be substantially less intuitive to the user, but more capable of handling the complexity of a larger network.

A second architectural advantage to a PAN structure is its frequency reuse or network capacity. If a single PAN cell supports 1 Gb and can be repeated in each room, the amount of network throughput in a home or office is reached by taking the number of rooms multiplied by the bandwidth available in each. By contrast, if one attempts to cover larger areas, the data rate for the individual cell is reduced and the number of cells in a building is also reduced. The total bandwidth supported in the building is reduced accordingly. To radio-savvy readers, this is a simplification of the situation that may not hold true under every condition, but which is essentially valid. However, the bottom line does hold well. If all other factors are constant, the use of smaller cells will tend to enable greater aggregate network throughput but at increased capital cost. This is increasingly important as the market continues to explore ever-increasing file sizes (think 80 GB portable music players) and streaming content (24 Mb/s video streams). One can readily imagine networks that are geographically concentrated to keep capital costs in check but which are very high throughput in order to address the transfer of very large blocks of data.

In practice, it will not be possible to have networks exist in isolation. A device operating on a PAN may need to communicate across the Internet. A LAN network may need to offload traffic to a PAN cluster to avoid saturation. Because of this, the network structures need to contemplate a somewhat more complex architecture, in which PANs connect to LANs and LANs to WANs.

As an example of this, consider a likely architecture for a free-standing home. In this environment, the first order of business is to gain connectivity throughout the home. This suggests a backbone LAN network that is capable of providing some level of communication throughout the home. Obviously, this could be provided by a variety of wired services (co-ax, power line, category 5 twisted pair) or it could be provided via wireless LAN. If the number of people in the home is limited and they tend to restrict themselves to email and web surfing, this is probably sufficient to get the home network done.

If, however, the family is larger or wishes to move streaming video across the network, it may be necessary to expand the network capacity by creating hot-spot PANs that can offload the backbone. Figure 2.4

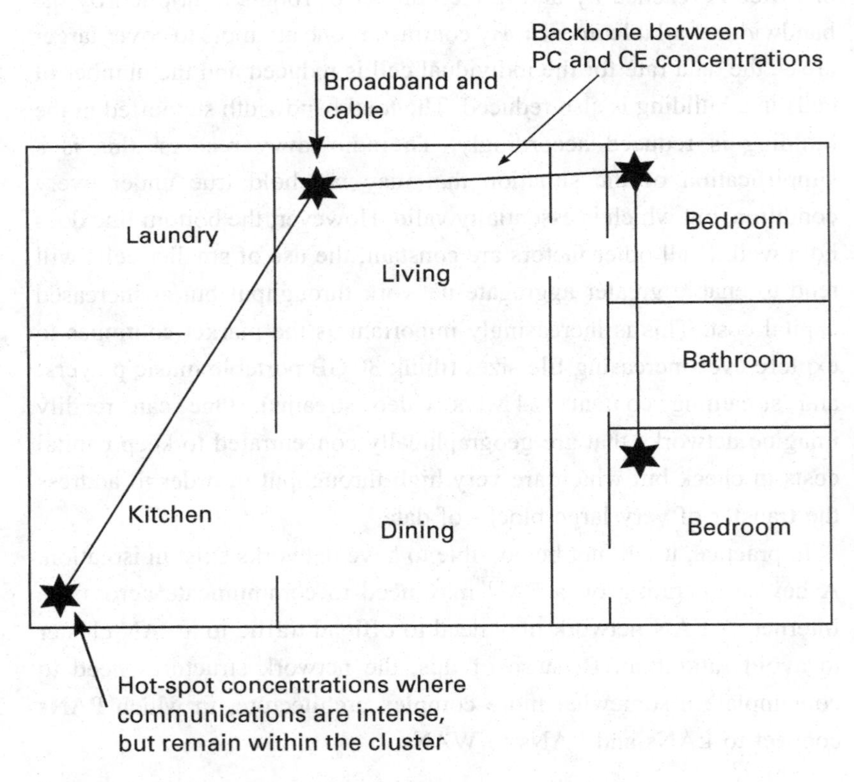

Figure 2.4 Backbone and hot-spot architecture

provides a picture of how this might be accomplished. For instance, there is a great deal of communication that takes place between the television, digital video recorders (DVRs), set-top boxes (STB), and other devices in the consumer electronics cluster. These devices usually exist within a single room and deal primarily with high-bandwidth streaming applications. The cluster or hot spot is represented by a star in the figure. It may be desirable to isolate that cluster in a PAN and then connect the PAN to the backbone network for traffic, which must cross the home or go out onto the Internet. This communication between clusters is shown by a line connecting the stars. Ultra-wideband PANs are being designed to perform this function. While UWB PANs are clearly optimized to operate in an isolated environment, it is possible to use them effectively to build a larger network structure when required.

2.6 Summary

While UWB was originally conceived of by regulators as being used for HDR, LDC and various imaging applications, this text is predominantly focused around HDR as it is believed to have the greatest potential for unit volume and market value and because both authors have an admitted bias in this direction. When looking at UWB's technical capabilities and comparing these with the range of HDR applications, several characteristics stand out. Ultra-wideband is able to achieve very high throughput, which is needed for large-file transfer and streaming-video applications. Ultra-wideband can be built at the very low price points that are necessary for consumer targeted products. The ability to resolve the location of UWB devices to within a few centimetres because of UWB's very sharp signal shape opens new applications which require location tracking. The very low power per bit transferred at which UWB is able to operate makes it possible to use the radio on battery-powered applications.

In addition to comparing UWB's technical capabilities with HDR applications, one should also look at the network architecture relative to these same applications. Ultra-wideband is a single-room technology, which is optimized for that purpose. Despite the inevitable tendency for companies to experiment with other approaches such as mesh networking and UWB as a LAN, these are sub-optimal implementations which will gradually be eliminated by the market.

References

[1] www.blu-raydisc.com/Section-13470/Section-14003/Section-14007/Index.html, accessed 4 October 2007.
[2] www.vesa.org/press/DP1.1pr.htm, accessed 4 October 2007.
[3] www.usb.org/press/WUSB_press/2007_07_23_USB-IF.pdf, accessed 4 October 2007.

3 Physical-layer (PHY) characteristics

The high-data-rate UWB products in development and shipping today are based on what is known as the WiMedia common radio platform (Figure 3.1). This chapter will provide a somewhat abbreviated technical overview of the physical layer, which makes up the lowest portion of the radio design. The intention here is to hit the highlights. For the reader who needs to understand the nuts and bolts of the radio's design, the ECMA-368 standards is recommended. [1] The content will focus on some of the more interesting facets of the design, which may be needed by individuals desiring a system view of the radio.

At the base of the common radio platform (Figure 3.2) lays the physical layer (PHY). Originally coined as a networking term, PHY refers to the combination of software and hardware programming that defines the electrical, mechanical and functional specifications to activate, maintain, and deactivate the transmission interfaces (or links) between communicating systems. The PHY may or may not include electromechanical devices, but in essence, it is the brains of the radio. Basically, the PHY's job is to transmit bits of data over a communication medium in either digital or analogue form. It makes no difference as to what those bits represent; the PHY operates in the same way regardless of the type of data. Physical-layer specifications typically define characteristics such as voltage levels, the timing of voltage changes, data rates, maximum transmission distances and physical connectors. They determine how data will be handled on the interface, transmitter or RF carrier.

This is also the case with the WiMedia PHY specification. The WiMedia PHY specification is laid out in the ECMA-368 standard, first released in December 2005 by Ecma International, the European association for standardizing information and communication systems (discussed in Chapter 5).

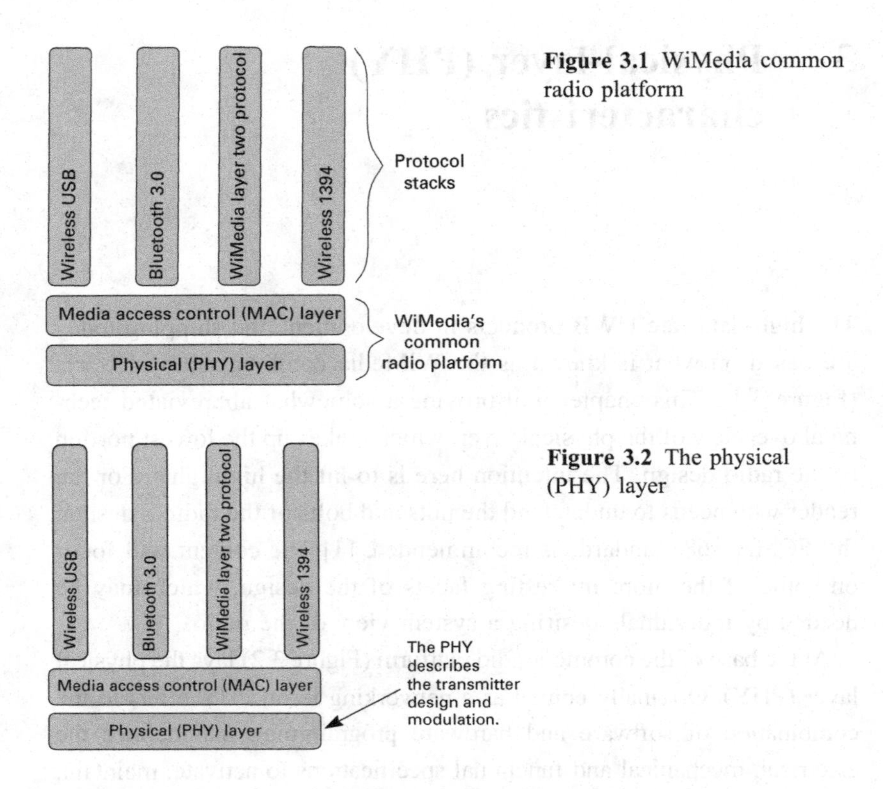

Figure 3.1 WiMedia common radio platform

Figure 3.2 The physical (PHY) layer

This Ecma standard specifies that the UWB PHY for a wireless personal area network (PAN) that uses the unlicensed 3100 to 10 600 MHz frequency band. It must support data rates of 53.3 Mb/s, 80 Mb/s, 106.7 Mb/s, 160 Mb/s, 200 Mb/s, 320 Mb/s, 400 Mb/s and 480 Mb/s. The specification divides the UWB spectrum into fourteen bands, each with a bandwidth of 528 MHz. The first twelve bands are grouped into four band groups, each consisting of three bands, and the last two bands are grouped into a fifth band group, as shown in Figure 3.6.

An example of a block diagram for a multi-band orthogonal frequency-division multiplexing (OFDM) receiver is illustrated in Figure 3.3. The received signal is amplified using a low-noise amplifier and down-converted in frequency. The complex base-band signal is low-pass filtered to reject out-of-band interferers. This is then sampled and quantized using a 528 MHz analogue-to-digital converter (ADC) to obtain the complex digital base-band signal.

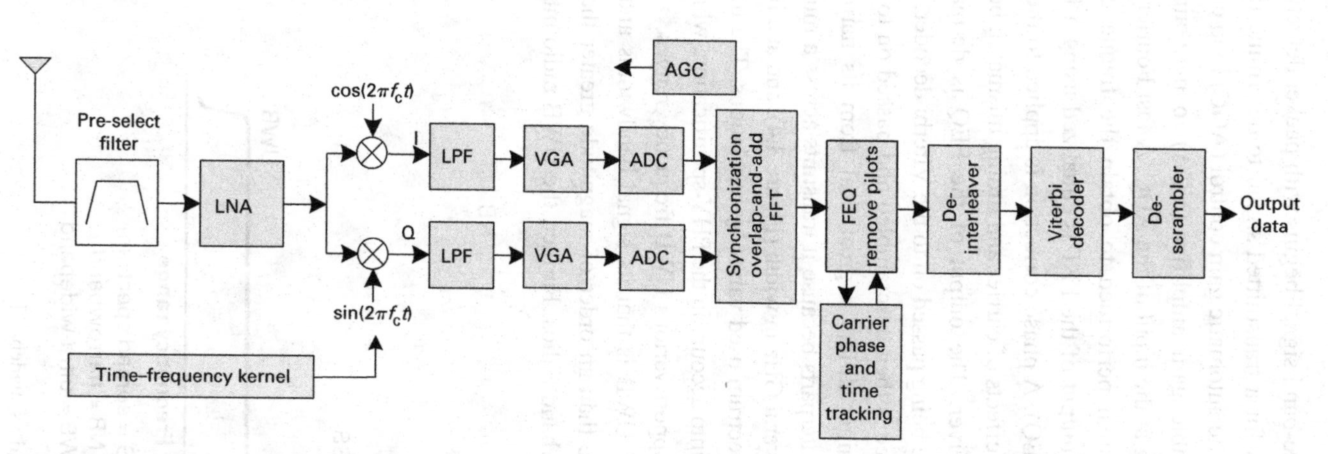

Figure 3.3 MB-OFDM block diagram

Processing the base-band signal begins with packet detection, where the receiver searches for a transmitted signal to determine if there are any packets on-air. The automatic gain control (AGC) loop controls the settings of the variable gain amplifier (VGA) to maintain the best possible signal swing at the input of the ADC. A fast Fourier transform (FFT) operation is then performed to obtain the frequency domain information and the output of the FFT is equalized using a frequency-domain equalizer (FEQ). A phase correction is applied to the output of the FEQ to undo the effects of carrier and timing mismatch between the transmitter and receiver. The output of the FEQ is de-mapped and de-interleaved before being passed on to the Viterbi decoder. The error-corrected bit sequence is then descrambled and passed on to the MAC.

The UWB transmitter differs a great deal from its narrowband or spread-spectrum counterparts because it transmits across a much broader spectrum, ranging several GHz in width (Figure 3.4), and at a much lower power. The actual spectrum used varies by regulations. The transmitter's needs must be taken into account in the PHY specification, which must be general enough to support various UWB frequency ranges.

One challenge for UWB is that its signal bandwidths and fractional bandwidths are more than an order of magnitude greater than those of most existing narrowband radios. Further, the UWB radio must co-exist

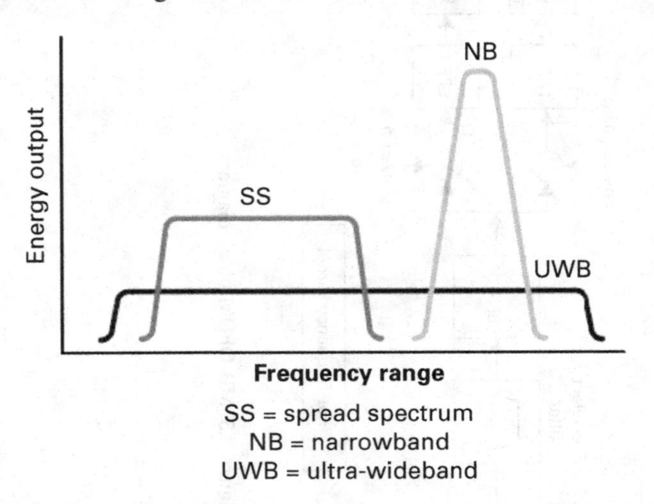

Figure 3.4 Narrowband vs. wideband

with numerous narrowband systems that are already transmitting and receiving at frequencies contained in the UWB system bandwidth (Figure 3.4). As a result, the UWB radio has different sensitivity and bandwidth requirements. This has resulted in radio circuit and architectural designs that are quite different from their narrowband counterparts.

Over the years, a number of technologies have been developed to take advantage of the UWB spectrum. Despite some early enthusiasm, impulse radio, where information is encoded in a pulse either in position, amplitude or polarity, has not gained traction in high-performance communications applications. Instead, high-performance UWB applications, such as Certified Wireless USB and Bluetooth 3.0, are being based on a multiband orthogonal frequency-division multiplexing (MB-OFDM) scheme to transmit information. This is the format embraced by the WiMedia Alliance, which, at the time of press had more than 300 members, including Intel, Samsung Electronics, Wisair, Staccato Communications, HP, Eastman Kodak, Sony, Texas Instruments, Microsoft, Alereon, Nokia Corporation, STMicroelectronics and NXP Semiconductors. The main reason for choosing an OFDM solution instead of impulse-radio architecture is that it has been shown to provide better throughput in realistic multipath scenarios and to allow for a lower-cost digital implementation for high-level integration.

3.1 Multiband

Although many companies first worked on and developed single-band impulse-radio technology, they often evolved to a multiband approach in order to address some technical and design challenges. For instance, they found that a multiband approach improved performance in the presence of interferers and made it possible to implement the radio using low-cost RF complementary metal-oxide semiconductor (CMOS) technology.

In a multiband approach, the UWB spectrum is divided into several bands (Figure 3.5), each with a minimum bandwidth of 500 MHz. These sub-bands may be transmitted in parallel or sequentially. The transmissions can then be received by separate receive paths or by one frequency-agile receiver.

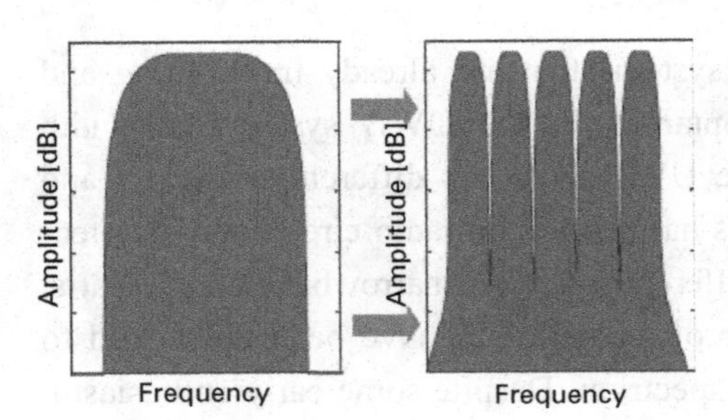

Figure 3.5 Ultra-wideband spectrum treated as one band or several bands

There are two types of multiband schemes, pulsed multiband and OFDM multiband, as explained below:

- Pulsed multiband transmissions modulate a specifically chosen pulse shape to obtain the frequency-domain properties for each sub-band. The signal is modulated in time or phase, using either pulse-positioning modulation (PPM) or binary phase-shift keying (BPSK). This approach realizes all of the multiband advantages mentioned above, but has some technical and performance drawbacks. Most notably, a pulsed system cannot address the issue of multipath fading using a single RF chain. A single receive chain is only capable of tracking one signal path. In a multipath environment, the energy of the transmitted signal arrives following multiple paths. For a pulsed system to capture this energy, it is necessary to create several RF chains. By contrast, an OFDM receiver is able to capture all of the signal paths using a single RF chain.
- In OFDM multiband, each band is further divided into subcarriers. The occupied bandwidth and spectral shape are largely defined by the inverse Fourier transform applied in the transmitter.

In all wireless systems, signals bounce off of items in the environment, and receivers inevitably detect multiple signals of varying strength and reliability. This phenomenon is referred to as multipath fading, or, more simply, multipath, and it is very likely to occur in an indoor environment. It happens because the transmitted signal divides and takes more than one path to the intended receiver. As a result, some signals arrive out of phase with others, cancelling each other out, resulting in a weak or fading signal. The capability of an RF signal chain to handle this multipath energy is essentially what determines the system's operational range. Different techniques can be employed to mitigate this effect, including separating the different paths, employing modulation techniques, coding techniques or receiver architectures.

3.2 Multiband orthogonal frequency-division multiplexing

The performance of a multiband system is improved with respect to a pulsed system when a different modulation scheme, such as OFDM is used. The Ecma PHY standard, which is used by WiMedia, specifies a multiband orthogonal frequency-division multiplexing (MB-OFDM) scheme to transmit information. This approach uses OFDM to convey the necessary information on each band, and then combines it with a multiband scheme to interleave the information across all of the bands in order to achieve the same power as an approach that simultaneously uses the entire bandwidth.

In addition to being able to efficiently capture multipath energy, an OFDM system also possesses several other desirable properties, including high spectral efficiency, inherent resilience to narrowband RF interference, and spectral flexibility – which is important because the regulatory rules for UWB devices have not been finalized throughout the entire world.

The transmitter and receiver architectures for a multiband OFDM system are similar to those used in a standard wireless OFDM system, as can be seen in Figure 3.3. However, the multiband part of MB-OFDM requires the use of additional circuitry to generate the hopping pattern for the transmission of each OFDM symbol because the RF signal must be shifted from different reference frequencies.

Figure 3.7 shows an example of how the OFDM symbols are transmitted across multiple bands.

For example, the first OFDM symbol is transmitted on sub-band 1, and the second one is sent using band 3. The third OFDM symbol is transmitted on band 2, and the fourth OFDM symbol is sent using band 1 again, and so on, as shown in Figure 3.7.

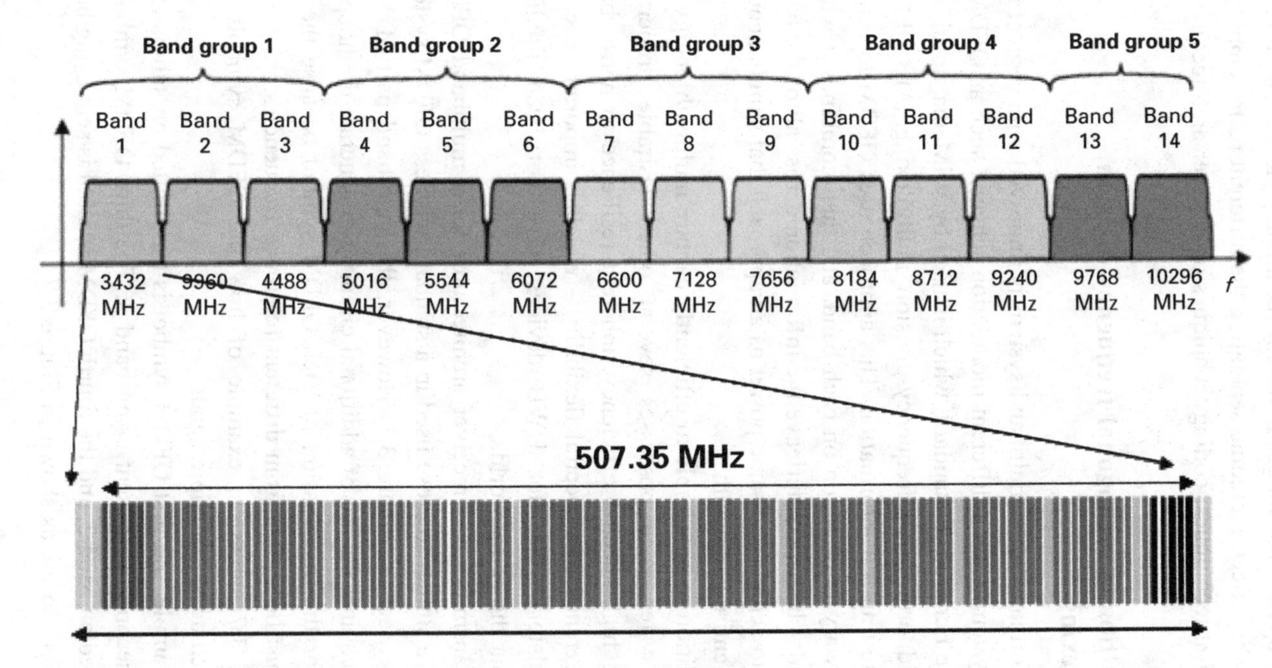

Figure 3.6 Ultra-wideband spectrum divided into bands and band groups

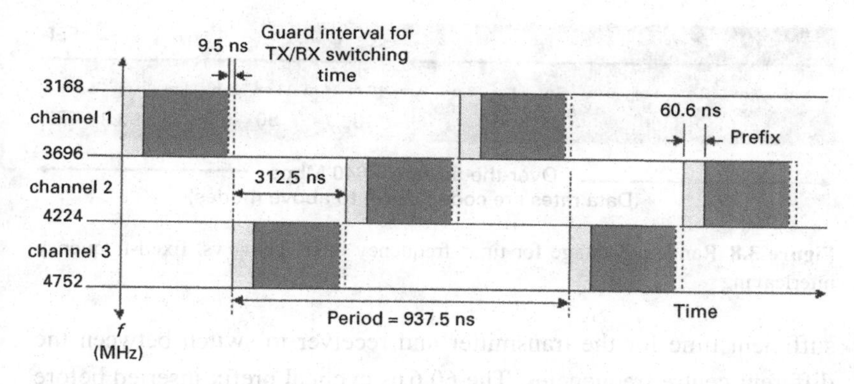

Figure 3.7 View of time–frequency coding for a multi-band OFDM system

Time–frequency code can also be defined in terms of length, such as length 3, 2 or 1. This means that the signal hops through 3, 2 or 1 bands. Time–frequency codes of length 3 provide a performance advantage in terms of frequency diversity and multiple access because they utilize more frequency; however, not all world regions have a spectrum allocation broad enough to allow all three bands and sometimes it is necessary to reduce the length of the code to 2 or 1.

The Ecma PHY standard requires that a UWB system support two types of time–frequency codes. When the coded information is interleaved over multiple bands, it is referred to as time–frequency interleaving (TFI); and, where the coded information is transmitted on a single band, it is referred to as fixed-frequency interleaving (FFI).

Fixed-frequency interleaving results in a shorter range; however, if the system is operating in an area that is dense with other networks, then it prevents interference, and the trade-off for range might be worthwhile. Time–frequency interleaving, on the other hand, is a frequency-hopping approach, and it offers a longer range (see Figure 3.8). The PHY must support both FFI and TFI operation.

Note that the symbols in Figure 3.7 also have guard intervals of approximately 9.5 ns (depicted by the dotted line following each symbol block). The guard interval is appended to each OFDM symbol so that only a single RF transmit and RF receive chain is needed for all channel environments and all data rates because it ensures that there is

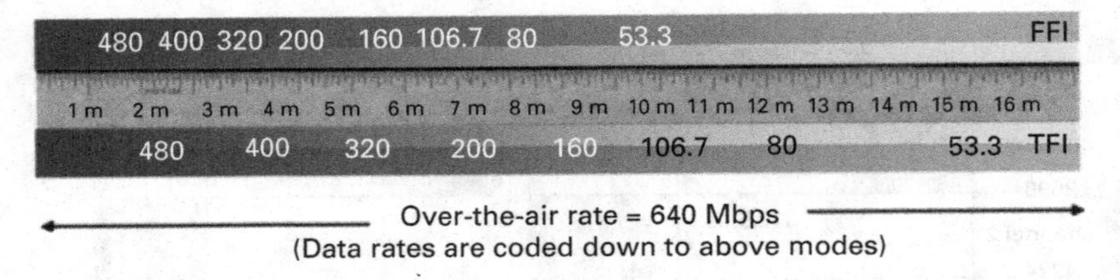

Figure 3.8 Range advantage for time–frequency interleaving vs. fixed-frequency interleaving

sufficient time for the transmitter and receiver to switch between the different centre frequencies. The 60.6 ns cyclical prefix inserted before each OFDM symbol helps alleviate the effects of multipath. The time-domain gap caused by the cyclical prefix between the transmitted and received signal allows the system to take into account multipath reflections.

3.3 Summary

The physical or PHY layer of the radio is the portion of the device that takes data given by upper layers and transmits the data over the air. If data flows in the reverse direction, the PHY is responsible for gathering the energy out of the air and converting it into data, to be passed onto upper layers of the stack to process. In the overwhelming number of HDR radios shipping today, the modulation design employed is the multi-band OFDM approach.

Instead of treating the entire available spectrum as a single contiguous block, MB-OFDM divides the spectrum up into a number of smaller sections. These divisions include band groups at the highest level, bands and finally sub-bands, which can be controlled individually or collectively to achieve various objectives. This approach to segmenting the spectrum is done to obtain channelization, to increase the radio's ability to respond to interferers and to provide flexibility for changing regulatory conditions.

The MB-OFDM uses the segmented spectrum through two principal schemes known as time–frequency interleaving, which uses three bands in varying sequences to transmit data, and fixed-frequency interleaving, which transmits data on a single band. These two techniques were originally designed to handle evolving US regulatory conditions, but also make it easier to build the radio in a CMOS chip process.

Reference

[1] www.ecma-international.org/publications/standards/Ecma-368.htm, accessed 4 October 2007.

4 Media-access control (MAC) layer

As with the PHY in the previous chapter, this discussion of the MAC is intended to be somewhat cursory. The full detail can be found in the ECMA 368 standard.[1] As demonstrated in Figure 4.1, the MAC layer sits immediately above the PHY.

The media access control (MAC) layer of the radio connects to the service access point (SAP) on top of the physical layer. The PHY SAP is nothing more than the logical gateway through which data flow in a specified format from the MAC to the PHY and back again. When data need to be communicated from one device to another, they must begin at the top of one of the protocol stacks shown above (WUSB, Bluetooth, etc.) and flow down through each layer, out over the connecting media (RF or wire) and up through each of the layers of the equivalent stack in the radio to whom the communication is being sent. Each layer of the stack has a specific task to perform in making sure that the data are successfully transferred.

Where the PHY is responsible for going through the physical steps of placing bits onto the air during a transmission effort and taking them off again during a receive operation, the MAC is responsible for the first level of processing that takes place on data coming out of the PHY.

For instance, a radio channel, by its nature, is relatively unreliable. Unlike a wire, which has very few transmission errors, a wireless environment has a comparatively greater number. The MAC is responsible for correcting these errors wherever possible. Sometimes this is done through retransmission requests. When retransmissions occur, the sequencing of the data is disturbed. Retransmitted packets are now in the wrong order relative to the way that they were originally sent into the radio. It is the responsibility of the MAC to correct for these situations. The MAC uses packet numbering schemes to restore the order.

Another critical MAC function is addressing. If the communication link or network is designed in such a manner that there are more than

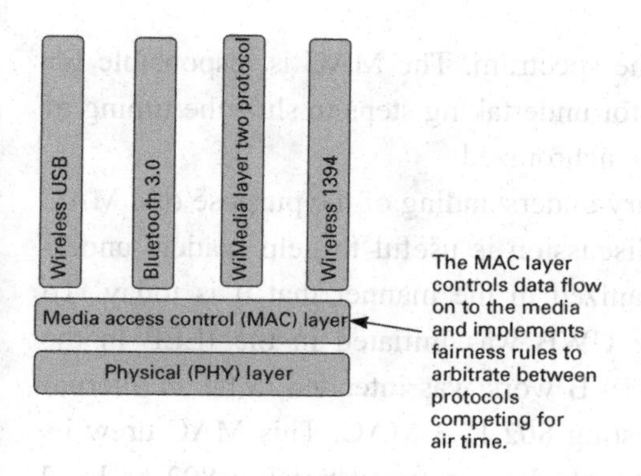

Figure 4.1 Media access control (MAC) layer

two communication partners, it is necessary for the transmitter to specify the receiver for which the communication is intended. This requires addressing. The MAC service access point (SAP) provides that unique address and inserts it into the communication flow as needed.

Once it is possible for data to be exchanged with the correct communications partner and the data are delivered intact and in order, the next function that the MAC must perform is managing contention for the medium. Say, for instance, that a video application is presently using the link but a second data application would like to do so as well. There must be rules that ensure that each application has access to the medium, but under terms that provide an acceptable level of performance to the consumer of the service. The protocols that are used to establish these guarantees are known collectively as quality of service (QoS). The UWB MAC employs several different protocols to address the needs of the applications that will use this radio.

As a final responsibility of the MAC, there are certain radio management issues, which it takes the MAC's intelligence to handle. Imagine, for instance, a situation wherein one network of UWB devices exists in a living room and a second network exists in an adjacent bedroom, which presently has the door closed, effectively separating the networks. Each network coordinates use of the spectrum independently by aligning its members around a synchronized timing of transmissions.

Now, imagine that the door between the two rooms has opened. The two networks can now hear each other and recognize that there is conflict between them for the spectrum. The MAC is responsible for detecting this condition and for undertaking steps to shift the timing of the networks until they are synchronized.

With this rather rudimentary understanding of the purpose of a MAC in mind, a brief historical discussion is useful to help readers understand why the MAC is organized in the manner that it is today. To begin, the original work on UWB was initiated in the IEEE in the 802.15.3a Committee. The UWB work was intended to be an alternative physical layer to an existing 802.15.3 MAC. This MAC drew its heritage from other 802 network designs (most notably 802.11 local area network designs). The 802.15.3 MAC employed a central control scheme in the same manner as 802.11's (WiFi) access point design.

As the work in 802.15.3a proceeded toward the selection of a physical layer, it became clear that a significant minority of participants were expressing reservations about the design of the existing MAC as well. The discontent was coming principally from consumer electronics (CE) manufacturers who believed that the existing design did a poor job at reflecting their application requirements. If UWB were to be successfully deployed into the CE markets, it was necessary for the standards work to address these concerns. Unfortunately, procedural limitations inside the IEEE made it either improbable that the work could be done or at least very challenging and time consuming.

When it became clear in 2004 that the IEEE process was unlikely to reach a successful conclusion to the PHY work (much less the necessary MAC changes), part of the compromise that was reached to get support for taking the work outside the IEEE was to modify the MAC design in such a manner that it was able to support a decentralized architecture. The initial development was done in a short lived organization called the Multi-Band OFDM Alliance (MBOA) which soon consolidated efforts with WiMedia and contributed their thoughts to the Ecma standardization process. While the process of merging SIGs appears rather convoluted, it was, in fact, a fairly efficient way to get the CE needs incorporated into a standardized MAC. This effectively changed the MAC design from the centralized 802.15 approach to a new decentralized architecture.

However, the centralized design was not completely dead. Shortly after the merge with the MBOA, WiMedia developed the concept of a common radio platform wherein the UWB radio design would be offered to other SIGs who had an interest in a high-rate radio. Most noteably, this was the Bluetooth and Universal Serial Bus groups (which will be discussed in detail in Chapter 6).

While both WiMedia and Bluetooth were generally comfortable with the decentralized model of the MAC, there was a problem with USB. Historically, the USB protocols were designed around the assumption that the PC could afford to be expensive and computationally dense, but that peripheral devices could not be. If the communications protocol required significant processing to be added to peripheral devices, this would be a non-starter. The peripheral devices could not raise prices to accommodate the radio.

Ironically, the solution to the problem turned out to be centralization. If a PC were able to act on the network as a peer, but would act as a proxy for the peripheral devices behind it, it would be possible to preserve the cheap peripheral logic that USB was founded upon. To do this, it would be necessary to modify the MAC to allow the PC to carve out a block of time on the network when it could communicate with its peripherals that would not conflict with the peer-to-peer operations.

In practice, this is implemented in two parts. First, the WiMedia MAC is designed as a peer-to-peer architecture. It does nothing explicitly to contemplate a centralized approach. However, it does have the ability to execute private reservations. This will be discussed later in the chapter, but for the limited purposes here, the private reservation is a window during which the device may use the spectrum in any way that it sees fit and using any protocol that it chooses. A device making the reservation can then use the spectrum to communicate with devices conforming to a second protocol. The USB uses this mechanism to establish the centralized architecture that it needs to achieve its design objectives.

The next portion of this chapter will refer separately to the WiMedia MAC and the WUSB MAC. It is technically inaccurate to refer to these as separate MACs, but it is a convenient way to describe the MAC level differences that exist between WiMedia and WUSB implementations. In referring to the WiMedia MAC, this includes the MAC elements that are part of the common radio platform and are employed by all users of that design. A reference to the WUSB MAC is intended to discuss only those mechanisms that are used by WUSB compliant nodes operating during a WiMedia private reservation period.

The following sections are intended to provide a brief description of the major operations that the WiMedia MAC is expected to perform. A more exhaustive discussion of the formatting of MAC transactions can be found in the ECMA-368 documents, which are available through Ecma International.

4.1 Channel selection

Once a WiMedia compliant UWB radio has powered up and is ready to begin operations, the first order of business is to find a channel on which to operate. A channel is comprised of a frequency band group which describes the frequency boundaries of operation, and a time–frequency code (TFC), which describes the manner in which hopping is conducted between the frequency bands that make up the band group. Once this information has been provided to the PHY, the station may begin to listen to the spectrum in search for somebody with whom to talk.

The WiMedia MAC does not establish a policy about where an application should begin looking for communication partners. While this sounds as if it has the potential to cause difficulty in establishing initial communications, it was done intentionally. Because WiMedia is designing a common radio platform, it is necessary to make the design general enough to be used by all of WiMedia's intended partner SIGs. These SIGs can and do have different preferences for the spectrum in which they choose to operate. Specifically, the Bluetooth SIG has a position wherein the UWB device is restricted to operation above 6 GHz. If the spectrum search algorithm were hard coded to reflect this position, it would require that searches begin at frequencies above 6 GHz. So far, so good, but there are two problems. First, both

WiMedia and WUSB intend to use the radio to communicate in a spectrum below 5 GHz. Second, this requirement to search the spectrum above 6 GHz is in conflict with another early-market reality. Radio designs above 6 GHz will be delayed relative to lower-band designs. They are simply not available at the present time. Instead, the MAC was described in a way to allow maximum flexibility. The upper-layer protocols (and their associated SIGs) will be responsible for defining how searching and other channel selection functions are conducted. The WiMedia MAC will be silent on that point.

So, for the moment, assume that the channel selection has been made by the appropriate body and the radio has now established communication with a piconet. If, because of channel degradation or DAA requirements (described in Chapter 8), it becomes necessary to change the channel selection after communications have begun, it is possible to move the members of the piconet in unison to a new location where no conflict exists.

The device that first detects the need to change channels simply issues a command to the other members of the piconet notifying them that a channel change is being advised and also notifying them of the recommended alternative location. Changing channels under these conditions is not mandatory for a radio receiving one of these commands, but it is expected that most devices will comply with the request if they are able to do so. Operations will then resume at the new location.

4.2 Beaconing and synchronization

Once the channel has been selected, the next step for the radio is to listen to the spectrum for transmissions by other UWB devices. If no other device is transmitting, the new device is free to begin transmitting a beacon for the purposes of setting up a network. Regulatory requirements allow the first device to beacon for a period of up to ten seconds. If a communication partner has not been found within that timeframe, beaconing must cease. This was done to prevent unnecessary signalling, which would only serve to raise the noise floor.

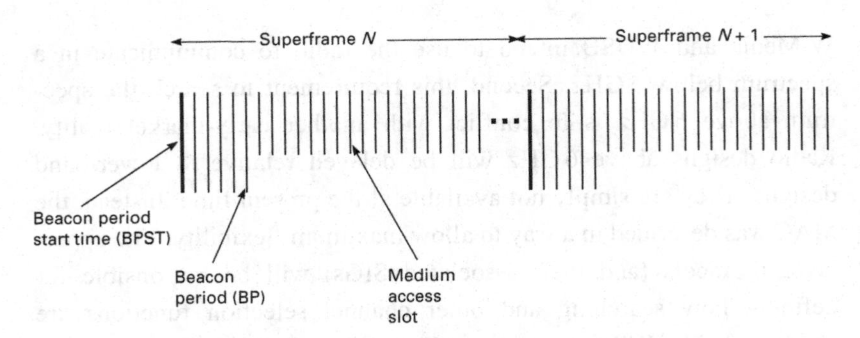

Figure 4.2 The WiMedia MAC superframe structure

If the new device begins listening and discovers that a network has already been established, the new device is required to join the existing network. To understand how this is accomplished, a quick description of the superframe structure and the beacon is in order.

The most fundamental component of Wimedia MAC communications is the superframe (Figure 4.2), which is the top-level-structure. All other commands and communications fit into it. The superframe is 65 536 microseconds long, and is divided into 256 media allocation slots (MAS), each 256 microseconds long. It repeats continuously. If one were to imagine a ski lift operating in a continuous loop to pick up and drop off information, this would be a reasonably close parallel. There are three major subdivisions of the superframe including the beacon period start time, the beacon slots and the media access slots. The beginning of the superframe is indicated by a beacon period start time (BPST). This point designates the start of the superframe loop. New devices wishing to participate in the network are required to synchronize their timing to coincide with the BPST so that the superframe timing is shared among all network members.

Once synchronization is established, the new device is required to listen to the existing superframe contents as the current network status is represented in the superframe beacon slots. Each active node is obligated to transmit a beacon in each superframe. By listening to the beacon portion of the superframe, a new device understands which nodes are nearby, what their capabilities are and what transmission slots have been reserved by other devices. With this information in hand, the new device then begins transmitting its beacon in the first available beacon slot that was not previously used by another node.

The beaconing process is used by the network members to maintain the organization of the network. The beacon portion of the superframe acts as a sort of communal bulletin board. Not only do devices post information about themselves, but they also post information about the neighbours that they can hear. By doing so, any individual node is able to create a map of devices up to two hops away, which can be used to coordinate activities. The two-hop approach also allows the network to avoid problems such as hidden node conflicts which might otherwise occur during medium contention periods.

4.3 Multi-rate support

The PHY layer supports throughput at data rates of 53.3 Mb/s, 80 Mb/s, 106.7 Mb/s, 160 Mb/s, 200 Mb/s, 320 Mb/s, 400 Mb/s, and 480 Mb/s. The MAC shifts between these rates at different times and for a number of reasons. For instance, the transmission rate during the beacon period is set at 53.3 Mb/s. This rate maximizes the range at which the beacon can be successfully received and so ensures the greatest integrity of the network.

The data rates are also negotiated between devices to provide reliable communication. In a manner similar to the logic used for the beacon data rate, one or both of the devices may choose to reduce the data rate of a link in order to minimize the packet errors. The data rate will also be used in conjunction with the transmit-power control features to provide a robust link at the minimum transmitted power.

4.4 Transmit-power control (TPC)

Devices that are compliant with the WiMedia MAC are required to operate using the least power necessary to maintain their links. To accomplish this, two devices who wish to communicate will employ a feedback loop. The recipient device may send an information exchange (IE) request to the sending device as part of its beacon to request a change in power.

The recipient device may use any one of a number of implementation-specific mechanisms to evaluate the quality of the link to trigger the request. This could include signal-to-noise ratio (SNR), received signal strength, error rates or other techniques. The implementers are free to use whatever method they choose to evaluate the quality of the link. The MAC standard does not attempt to limit the implementer in this aspect.

The transmit-power control feature of the MAC can significantly reduce the interference from UWB to licensed devices. When applications were evaluated in both the ITU and the European regulatory processes for their probable separation distance, a significant percentage were found to operate in the one-to-three-metre range. The UWB radio's ten-metre link capacity may well prove to be overkill in these cases. By reducing the emitted power in these applications, the interference generated is reduced. This becomes particularly material in cases such as fixed wireless access (FWA), where the FWA device has the potential to be in very close proximity to the UWB transmitter.

4.5 Power management

A significant percentage of the applications envisioned for UWB are targeted at battery-powered devices, such as smart phones, handheld computers or personal video players. In these cases, it is very important for the MAC to take steps to reduce power consumption whenever possible to maximize battery life. The TPC, described above, is useful for this purpose. In addition, the MAC has made provision for inactive devices to hibernate, in order to save power.

If a device decides that it is inactive and wishes to turn off its radio for a fixed period of time and move to a hibernation state, it is possible to notify the remainder of the network participants of this fact through a hibernation mode information-exchange notice in the beacon and then to cease further transmission. The hibernating device is required to post its anticipated return time as part of the hibernation message.

An analysis of the applications that are expected to employ UWB strongly suggests that the network traffic will be very bursty and intermittent in nature with significant periods of silence between the bursts (in most cases). This is most particularly true of file-transfer applications and transaction-oriented exchanges, where a person is involved. It is much less true of streaming video applications, where the traffic is much more regular and persistent.

As the transmission rates of UWB devices increase to levels of 1 Gb/s and beyond, the quiet periods can be expected to increase. Less time will be required to transfer a given file, resulting in more lengthy periods of quiet. The hibernation mode allows devices to take advantage of these quiet periods to shut down and save power.

While hibernation has clear advantages for power savings, it also has the risk of destabilizing the network. If too many devices were to hibernate simultaneously, it is possible for the network to become too sparsely populated and, potentially, to cease operations. When nodes return from hibernation, they would need to re-establish the network from scratch. But this is an avoidable condition.

Some devices in the network may also volunteer to act as hibernation anchors. This is to say that these devices anticipate remaining on the air and available. They will continue to maintain the status of hibernating neighbours in an effort to keep the piconet stable during hibernation. Devices that act as communication gateways or crossroads and are more likely to be connected to AC power will tend to take this responsibility in practice. In this way, the beacon information is preserved so that mobile devices can enter and leave the hibernating state without causing additional overhead or disruption to the piconet.

4.6 Range measurement

In prior discussions on applications, it was mentioned that range-dependent applications were relatively few in the HDR community. But they are not absent. The Wimedia MAC has been designed to include an option for range determination for those manufacturers who wish to implement applications requiring it.

Before range estimation between two devices may be undertaken, it must first be ascertained that both devices support the option. This information is found in the device's beacon fields of the superframe.

If both devices are found to support the option, it is possible to proceed with the measurement.

To obtain a ranging measurement, the concept is to have an initiator send out some form of a logical ping to a recipient. The recipient returns the ping. The initiator measures the round-trip time and uses the elapsed time to calculate the distance between the two devices.

But there are several inefficiencies that need to be accounted for to make sure that the only thing being measured is the travel time. There is a delay at the transmitter end. Once the counter has begun, some time elapses before the signal is launched from the antenna. This value has to be incorporated by the initiator in the calculation. On the receive side, since this is a logical ping (data) as opposed to a physical ping, a delay is inserted at the recipient end when it processes the request and then responds to it. To deal with this delay, the recipient needs to measure and respond with the amount of delay it has added to the equation.

The response from the recipient, combined with the data that are measured or estimated at the initiator can then be used to isolate the travel time and, therefore, estimate the separation distance. For greater accuracy, it is possible to take a number of readings in quick succession and average the results.

4.7 Bandwidth reservations

Because the WiMedia MAC was designed to handle large-file transfers, streaming video and a variety of other transfer types, two principal access mechanisms were created to support optimal transfer of data. These include prioritized contention access and distributed-reservation-protocol reservations. The contention-based access protocols were designed to work efficiently for transaction-oriented applications, while the distributed reservation protocol was designed to address the need for streaming applications to have dependable delivery timing and bandwidth.

4.7.1 Prioritized contention access (PCA)

The PCA protocol allows a device to contend with other devices in the piconet for access to the medium. Contention occurs for any unreserved media access slots or during a reservation which has been declared to be open for contention (soft reservation, described later in this chapter). For those individuals who are familiar with wireless LAN protocols, this approach should not be new. The PCA approach is based upon the IEEE 802.11e Enhanced Distributed Channel-Access Mechanism.

There are four classes of traffic, which are given priority in the following order; background, best effort, video and voice, with background having lowest priority and voice having highest. In this way, traffic which has time-sensitive payloads is given higher priority access to the medium relative to traffic that is less time sensitive.

The PCA protocol also implements mechanisms to assist with collision avoidance. Conceptually, a device attempting to gain access to the medium begins by listening for activity. Once it has been determined that there is no other device using the medium, access is attempted. If two devices collide on their attempt to gain access, both devices will back off for a randomized period of time (number of frames). The back-off interval is a function of the traffic priority.

The PCA also has a power-savings mechanism built in to minimize the amount of listening that is required during the contention process (the receiver makes up a material portion of the power budget). The device which has control of the medium is required to indicate the number of frames which remain to be transferred in the network allocation vector (NAV). When the NAV decreases to zero, the medium is available. By knowing how long the medium will be busy, it is possible to stop listening until it becomes relevant again.

At this point in the transmission sequence, it is possible that several devices may have collected data that they wish to transmit and are now waiting to use the medium. With this state of affairs, a collision is possible. However, the PCA protocol describes access queues, which can be used to prevent this form of collision.

A queue is maintained for each traffic class. The queue information is posted as part of the beacon process and is updated by each device as changes occur. Two devices with the same priority resolve their access conflict when they enter the queue. In the event that a device from a higher-priority queue is contending for access with a lower-priority device, the higher-priority device wins. No physical collision occurs. This situation is referred to as an internal collision.

4.7.2 Distributed reservation protocol (DRP)

The distributed reservation protocol (DRP) is the second major mechanism described in the WiMedia MAC. It allows for contention-free access to the medium. In this process, the beaconing period is used extensively to create and manage reservations for use of the medium. A device will make a reservation request to either an individual address or through a multicast. In this request, the desired MAS slots that the device wishes to occupy are named. More MAS slots are requested to obtain a fatter logical pipe. Selection of specific MAS slots also affects the latency of medium access.

One of the key rules in the DRP approach is that each device is responsible for defending the reservations of its neighbours. That is to say that if a device receives a request for use of a slot that has been previously granted to one of the device's neighbours, it is likely that a hidden-node condition exists and the device has an obligation to speak on behalf of the neighbour. The receiving device will decline the reservation request. When a reservation must be declined, a reason code must be offered, which describes why. This gives the requesting device more insight into the problem that was encountered. A MAS availability map is also returned to the requesting device so that it has a picture of the available MAS slots as seen by the rejecting node.

When a device has successfully requested and received permission to use a number of MAS slots for a reservation, this is referred to as a 'hard reservation' and it comes along with fairly strict use conditions. Only the reservation owner is permitted to use the time slot. If the reservation owner decides that the bandwidth will not be used and wishes to release it to others, it is necessary for it to send an unused DRP reservation announcement to other network nodes and to receive an unused DRP reservation response in return.

In addition to the hard reservation, there is also a soft reservation within the DRP system. This is a hybrid of the DRP and PCA techniques. The MAS slots are open for contention in a manner equivalent to the PCA techniques, but the soft-reservation owner is given priority access whenever he contends.

4.8 Co-existence of different protocols

There is likely to be a wide diversity of UWB wireless multimedia devices and protocols, including Internet protocol (IP), Wireless 1394, CWUSB, and Bluetooth. None of these protocols will operate directly over the UWB radio, but each one must be adapted with an appropriate software layer. This is typically done using a protocol adaptation layer (PAL) (see Figure 4.3). The WiMedia MAC includes a component that provides a common MAC environment to interface with the PALs. (Note that a single application could have multiple PALs, and, in these cases, the MAC specifies how to coordinate the shared access to the medium.)

In effect, the MAC specification defines data formats, facilities and methods that will enable multimedia and other applications to share the resources and functionality of UWB wireless personal area networks (WPANs). In general, the specification addresses provisions for multiple application protocols, device and WPAN management, quality of service (QoS) that is appropriate to the type of data, secure authentication, interference mitigation between nearby devices and WPANs, collaborative time-channel use, device and protocol discovery, and power management.

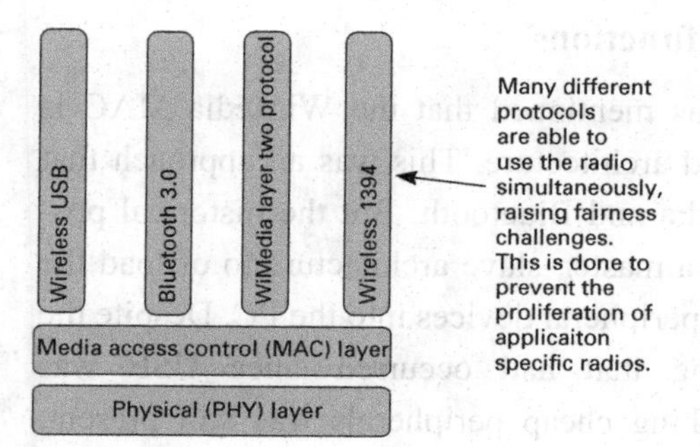

Figure 4.3 Protocols employing the UWB radio

Perhaps the most notable challenge of allowing multiple PALs to coexist in the same medium is to develop a fairness policy, which regulates reservations and bandwidth usage within the WPAN. This policy is designed to allow CWUSB, Bluetooth and ad-hoc networking all on the same MAC to share the same space. The fairness policies are incorporated into the MAC's reservation protocols.

The inherent high-bandwidth of UWB and fine time granularity of the WiMedia MAC allow trade-offs to be made so that systems based on WiMedia can provide high bandwidth and low latency when seen from a human scale. For example, USB mice get polled every 8 milliseconds. While this is a short time span on the human scale, this represents 32 MAS, which translates to a long time on the WiMedia time scale.

The implementation of these reservation schemes is left up to the individual company building devices. Interoperability is well defined, but some companies will achieve better throughput than others. Specifically, there are two aspects of the fairness policy in the MAC specification – both concern reservation usage. The first rule is known as the safe medium-access slot(s) rule, which basically says that no device can reserve more than half the available bandwidth as defined by the MAC. This rule specifies that any unused MAS have to be given up if another device asks for them. The second reservation rule states that if a device makes a reservation, then it must use it. The rules specify a certain percentage of the reserved time that must be used for transmission. If the reservation is underused, it may be shortened.

4.9 Wireless USB MAC functions

Earlier in this chapter, it was mentioned that the WiMedia MAC is designed around a distributed architecture. This was an approach that was adopted both by WiMedia and Bluetooth. But the historical position of USB was to employ a master–slave architecture to offload the processing expense from the peripheral devices into the PC. Despite the reduced costs in processing that had occurred since USB was developed, the value in making cheap peripherals was still present.

Because of this, the master–slave model needed to be enabled in the architecture of the WUSB MAC.

By using the private reservation option that was built into the WiMedia MAC, it was possible to have an extremely high degree of flexibility in the design of the WUSB MAC functions. The private reservation essentially allowed WUSB to design a world within a world. The reservation allowed the PC to communicate with peripherals using almost any protocol and format it required.

As the WUSB engineers looked at this problem, one of their tasks was to preserve as much of the existing wired USB protocol as possible in the transfer. If the existing mechanisms were preserved, it would be possible to develop components more expeditiously and at lower cost. However, the move from a wired environment to a wireless domain forced the designers to make certain accommodations.

A wired USB device is never out of range nor does the term 'hidden node' have any meaning. Wired peripherals need never worry about seeing traffic from multiple hosts. With a wired connection, there is almost no chance of unauthorized data modification. Eavesdropping on a wired USB link is beyond the capabilities of most non-governmental entities. And finally, the raw error rate that is present on a wireless link is substantially higher than a wired connection. All of these factors forced designers materially to redesign the operation of USB to work in a wireless environment.

4.9.1 Wireless USB addressing

The USB devices are assumed to be hosts (PCs) talking to peripherals (devices). From an addressing perspective, this reduces the complexity of the task involved. Instead of connecting to an almost unlimited number of devices around the planet, the host needs only to talk to a small number of devices within a very short distance. Because of this, the USB addressing scheme is substantially different from the approach used in the WiMedia MAC.

The WUSB supports 127 unique addresses, which are all managed locally by the host. Additionally, specific addresses have meaning to the USB world. Address 255 is given to devices who are on the bus but are in an unconnected state. Under normal conditions, this state is a precursor to a request for connection. The range of addresses from 128–254 is used to initiate a clusterwide communication and is also used to communicate to a device before it has been authenticated.

4.9.2 Host channel

The WUSB host channel is comprised of the communication that takes place between a host and its devices. At the highest level, there are two parts to the communication. The first part is a control frame known as the microscheduled management command (MMC). The MMC is where the host will describe the transaction that is to follow. Following the MMC is the period in which the transaction is to be conducted. The MMC also states the time at which the next MMC will begin and thereby implicitly defines the conclusion of the transaction period.

4.10 Summary

The MAC layer provides a wide range of functions for the radio, covering topics such as addressing, error recovery, packet sequencing, channel selection and management, and media management. If one were to imagine the responsibilities being placed into a sequence order, they would start by having a new radio select a channel on which it would like to operate. This then proceeds to a listening and capabilities exchange protocol, which is conducted through beaconing. Once the radio has joined the network, it is then able to request use of the spectrum. This is done through one of the reservation processes, which make sure that spectrum use is conducted in an orderly manner. When data are transferred, the MAC is responsible for applying addresses to outbound data and reading data that are intended for its address. If received data are in error, the MAC is responsible for requesting retransmission and re-ordering the corrected packets when they are available.

Because the WiMedia MAC is designed to support a number of different applications, it was necessary to design flexibility into the

MAC to accommodate their needs. The most noteable accommodation was in wireless USB. Where the MAC is generally structured to allow peer-to-peer communications, the WUSB application needed to accommodate peripheral devices with very limited intelligence. Because of this, the WiMedia MAC allows a PC to perform a proxy function for WUSB peripherals. The PC operates as a WiMedia peer, but then reserves time in which it may communicate with its associated peripherals using its own communications protocol. Through mechanisms such as this, the WiMedia MAC has become a flexible platform that can accommodate a number of different protocols, which will be described in the following chapters.

Reference

[1] www.ecma-international.org/publications/standards/Ecma-368.htm, accessed 4 October 2007.

5 Implementation information

The information that one finds in the standards and specifications of a technology are frequently only part of the story. While standardization is critical to achieve industry-wide interoperability, it is also very important for individual manufacturers to be able to differentiate their products. For this reason, standardization bodies generally restrict themselves to describing those elements that are absolutely required to establish interoperability or common customer experience and usually remain silent about the rest of a design.

In the case of UWB, there are several points that are not described in the standards, but which one might wish to be aware of. For instance, there are many cases in which it will be necessary to place a UWB radio alongside one or more other radios as part of a general system. Some effort is required to get these devices to co-exist. There are also trade-offs that a designer will need to make on topics such as the level of integration that is desirable, the chip-packaging trade-offs and the antenna-selection options. Each of these issues is discussed in the sections which follow.

5.1 Co-location with other radios on the same platform

Because it is a wireless technology, UWB is subject to more scrutiny in terms of interference – including its effects on neighbouring devices as well as their effects on it. This is to reduce negative effects both to other UWB receivers and non-UWB receivers. In addition to the reference models for MAC restrictions, the WiMedia specifications also provide a set of conformance requirements based on the UWB device's over-the-air performance.

It is generally known that when several RF devices operate independently within close proximity to each other, some device transmissions might interfere with reception by other devices. To develop mitigation policies to reduce negative effects, it is necessary to understand the causes and nature of the interference. It is important to note that interference can occur with any MAC or PHY. Depending upon the application and circumstances, some platforms might require more complex mitigation techniques. The WiMedia specification addresses interference issues by defining some basic mitigation policies.

To understand the potential interference problems, consider this example: a set-top box (STB) capable of transmitting a video stream to a nearby high-definition television (TV). Figure 5.1 illustrates the independent operation of these two devices, meaning that no other RF devices are in the vicinity.

Admittedly, this is a simplified view of an actual RF environment, but in Figure 5.1 the set-top box's transmissions are shown by concentric arcs. The colour gradient – from black to light grey – represents signal attenuation. The dashed circle indicates the limit of interference created by the transmissions. If another RF device were located within the circle, its reception from sources other than the set-top box might be negatively impacted.

A major factor in how much interference a device will experience is the power level of the potential interferer. It is important to understand that not all devices transmit at the same power level and that transmissions occur in three dimensions (hence, interferers can come from all three dimensions as well). In addition, all devices are not equally

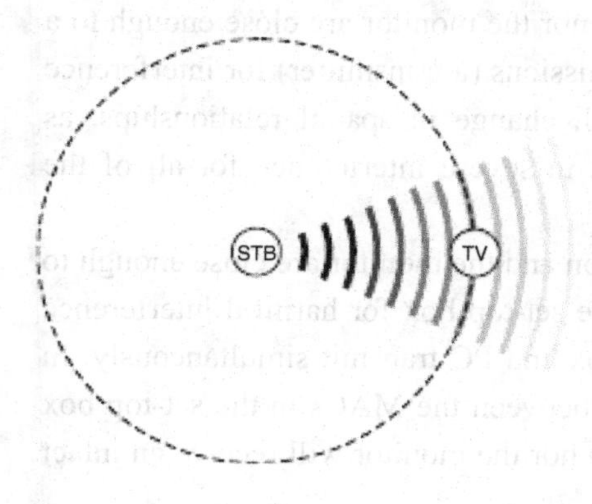

Figure 5.1 Simple communication link

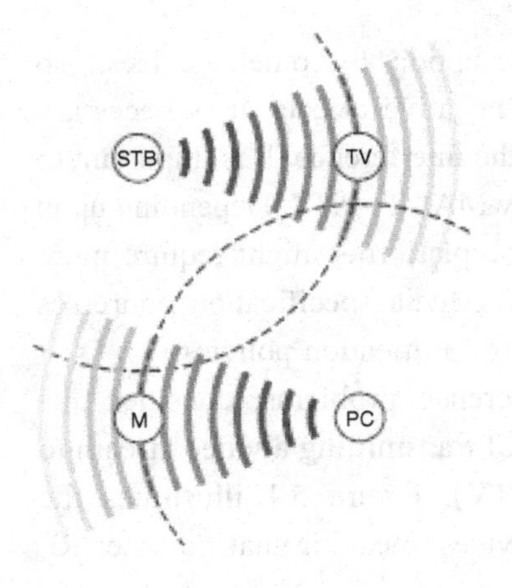

Figure 5.2 Several devices without interference

susceptible to interference. Their susceptibility is determined by the robustness of their data formats as well as their receiver sensitivity (which is a measure of the receiver's ability to detect very small or low-powered signals).

When RF devices operate in isolation, as in Figure 5.1 above, there is no interference, so they can use all of the channel time. If more devices are added to the scenario, as in Figure 5.2, their spatial relationship determines whether it is possible for them to operate independently without interference. It is in these and the following scenarios that the robustness of the MAC and PHY are tested.

In Figure 5.2, a personal computer (PC) and a wireless display monitor (M) have been added to the layout in the previous scenario. Although all four devices are shown in relatively close proximity to each other, neither the television nor the monitor are close enough to a competing source of radio transmissions (a transmitter) for interference to occur. However, only a small change in spatial relationships, as shown by Figure 5.3, can result in severe interference for all of the devices.

In Figure 5.3, both the television and the monitor are close enough to the transmitters in the PC and the set-top box for harmful interference to occur whenever the set-top box and PC transmit simultaneously. In the absence of any coordination between the MACs in the set-top box and the PC, neither the television nor the monitor will receive an intact

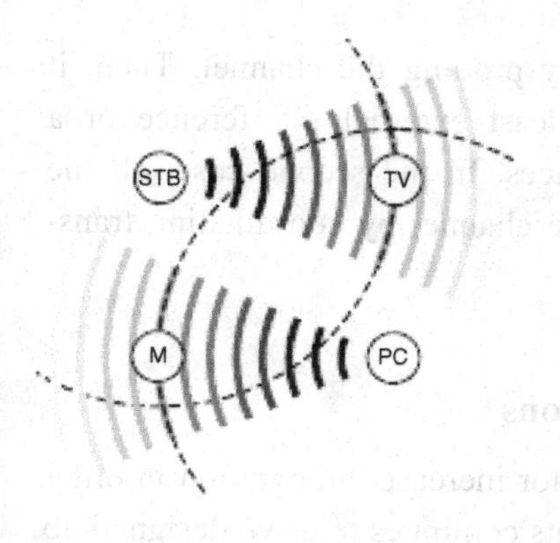

Figure 5.3 Several devices with interference

video stream, because they will transmit at the same time. These are simplified examples of static UWB implementations, but dynamic scenarios exist as well. The market for mobile devices continues to grow, and many individuals move around homes and offices with multiple mobile devices. For instance, what would occur if a user on the WPAN walked into the room with a personal digital assistant (PDA) that must synchronize with the PC? Typically, PDAs transmit with a low power level; however, the TV in the example is fairly close to the PC, and interference could be a factor.

Analysis suggests the following considerations when designing a layout that includes multiple devices on the same WPAN platform:

- Independent operation (Figure 5.1) is preferred whenever possible, since it permits maximum throughput and minimum latency.
- Devices can be co-located without causing interference, depending on their spatial relationship (Figure 5.2).
- As the distance between independently operating devices shrinks, a point is reached at which one or more of the devices is significantly impacted by interference (Figure 5.3).
- Interference is possible from mobile RF devices moving into the space.

It is important to understand that each RF device experiences the local environment from its particular spatial, temporal and performance point of view. And, this is very likely to differ from those of other RF devices. Before initiating transmit operations, a WiMedia-conformant device assesses channel conditions by probing the channel. Then, it either selects the channel with the least external interference or a channel already in use by other devices. In the second case, all the devices collaborate in their use of the channel by coordinating transmissions at different times.

5.2 Chip-integration considerations

In all wireless applications, the need for increased integration in order to meet cost, size and power constraints continues to drive designers to find new techniques and methods to achieve these reductions in components and board real estate. The need for high levels of integration in a single chip and advanced packaging solutions continues, and UWB is no exception in its place on the integration roadmap.

Packaging is critical to the success of a device, and must address issues such as high-frequency loss, parasitics, radiation susceptibility and discontinuity capacitances. All of these conditions can quickly degrade the performance of the best designed devices. And, for an emerging technology such as UWB, cost is crucial.

5.2.1 Integration

One way to achieve low cost is through integration, and the drive towards single-chip CMOS (complementary metal-oxide silicon) solutions is serious. A single-chip highly integrated approach eliminates the interconnects between multiple chips, and it can significantly improve performance, reliability and the need for board real estate. An IC that encompasses both the wireless RF radio and baseband functions can certainly improve cost, performance and reliability.

CMOS brings many benefits, the foremost being single-chip integration. This is because CMOS can deliver both the digital baseband processing and the high-frequency analogue radio in the same physical silicon (system memory and flash memory can also be integrated). In addition, because CMOS is lower cost than competing silicon–germanium (SiGe) semiconductor processes, it offers the potential to deliver a complete system-on-a-chip at a price-point appropriate for high-volume consumer-electronics offerings. Pushing the technology into the digital domain reduces any of CMOS's comparative disadvantages, most notably the difficulty associated with integrating RF-components in submicron CMOS processes. An exemplary precedent has already been set as CMOS radios dominate the Bluetooth market. Furthermore, 802.11a/b/g radios have also been implemented in CMOS.

While the production of UWB systems using CMOS technology is a high priority for most consumer electronics manufacturers (the milestone of a production CMOS radio is widely regarded as enabling the high-volume market), it does create some challenges for chipmakers. Ultra-wideband technology is centred on fast-changing wide-bandwidth signals, which necessitate high-speed circuits. Generating these wide-bandwidth signals at the transmitter and sampling them at the receiver can be a task. To receive a 500 MHz bandwidth signal, a baseband analogue-to-digital converter (ADC) is required to sample at 500 Msps, minimum. This is quite a task for today's ADCs.

For maximum integration opportunities, designers always turn first to CMOS, which makes it possible to produce highly integrated low-cost devices. Other silicon technologies, such as SiGe or BiCMOS, require the use of more specialized plants and typically utilize smaller wafers, which keep costs higher and production volumes lower even when in full production. For years, BiCMOS technology (bipolar and CMOS devices integrated on the same silicon wafer) was the prime choice for wireless devices, most notably, cordless and cellular phones. Recently, the tables have turned with the integration efforts favouring CMOS over BiCMOS. Most of the reasons centre on cost factors.

With CMOS, transceivers can be integrated with the baseband and RF functions on a single, low-cost CMOS chip, which is not only smaller but also inherently less expensive. Nowadays, CMOS integrated circuits include complementary silicon components, such as a baseband, MAC, high-performance input/output (I/O) memory, as well as other technology protocols and functions. The only inhibiting factor now appear to be that only a handful of companies truly understand how to provide a complete CMOS solution.

In the drive for integration, the UWB industry is no different from other wireless technologies. The industry is leaning towards all-CMOS solutions, in large part because MB-OFDM technology is implementable in high-performance, low-cost standard CMOS.

5.2.2 Packaging

Package-type selection is dependent upon three key factors: application, budget and size goals. A UWB IC can be implemented either as a packaged part with the motherboard supporting the required components that are not part of the IC, or as a module.

If implemented as a packaged part, the package must be capable of sustaining high-frequency performance. The key is maintaining short wirebond and lead lengths to minimize inductance, as well as to provide a very good ground for the RFIC through an exposed die flag.

For modular applications, the package style generally falls into two categories: land-grid array (LGA) or ball-grid array (BGA). It is important to understand that the package dimension and type will affect the product's cost. The primary difference between the LGA and the BGA is in the method of connection to the motherboard. The LGA utilizes land areas for the input and output connections that are soldered to the motherboard, while solder balls are configured in an array fashion for the BGA. The use of one over the other is likely to be driven by the number of connections.

While laminate implementations have been demonstrated, packages can also be made with a variety of other substrates, including ceramics, glass and silicon. Low-temperature co-fired ceramics (LTCCs) are being used today for Bluetooth and UWB modules. Glass is beginning to be used as well. Ceramic circuits can be realized both as LGAs and as BGAs and, like glass and silicon, they can also be configured as inverted BGAs. (Inverted BGAs are generally required for glass and silicon modules owing to the difficulty of making holes through the substrate.)

Many communications designers are looking at low-temperature co-fired ceramic (LTCC) solutions to meet the tough technical demands of their packaging and system requirements. These LTCC packaging materials have appealing RF properties, including superior multilayer capability, which allows designers to pack more functionality into a device, and provide dimensional stability and low dielectric losses for better performance. The process also boasts good repeatability, which improves performance and production yield. All of these features translate to real benefits for high-frequency application, such as UWB.

For UWB, integration of front-end filters, baluns, discrete microwave components, like inductors and capacitors, and low-loss transmission lines into the packaging-mask set reduces the bill of materials (BOM) and overall cost. The reduction in parasitics associated with bond wires and the optimized placement of critical RF interconnects can save the end user substantial problems in integration while improving overall performance.

The multilayer capability of LTCC allows integration of functions that were traditionally handled by discrete components, and this provides reductions in package footprints. One particular challenge with RF, especially when combining digital and analogue functions in a single package, is the need for isolation in the RF portion of the circuit. By taking advantage of the ground planes between layers and vertical division via walls, chip designers can still provide the necessary RF isolation within the package not only for RF-to-RF interaction but also, and perhaps more critically, to isolate the sensitive RF circuitry from the noise generated by the nearby digital circuitry.

5.3 Antenna considerations

Antennae that span ultra-wide ranges of spectrum were successfully designed decades ago. In many cases, they have the bandwidth and size to handle commercial UWB applications today. However, most of these antennae were designed for wideband military pulsed applications, such as radar, and are not feasible to be used in commercial application, owing to their size and cost. One might even consider using an amplitude modulated (AM) antenna as a UWB, since it covers 535–1705 kHz for a fractional bandwidth in excess of 100%. Commercial UWB, however, differs significantly from these applications in its operation because the UWB signal has specific requirements in terms of radiative emission.

An explanation of antenna theory is beyond the scope of this book. This section aims to explain the top-level considerations for a UWB antenna, with a specific focus on small, low-gain antennae that would be well suited for mobile and portable UWB devices.

An antenna converts electromagnetic energy that is travelling in a transmission line into radiated electromagnetic energy in free space, and vice versa. Growing global interest in the commercial deployment of UWB systems has encouraged the development of new ultra-wideband antennae. Because the FCC and other regulatory bodies have specified maximum power levels, every decibel (dB) counts in the radio system, and the quality of the antenna could make or break a UWB system design.

Traditional UWB antennae, such as those used for radar or AM radio, are actually designed to function as multiple narrowband antennae rather than having the ability to receive a single signal across their entire operating bandwidth. A commercial UWB system needs to have an antenna that can receive all frequencies simultaneously. As a result, the antenna's behaviour and performance has to be consistent and stable across the entire band.

For the best performance, a UWB antenna should be non-dispersive, which means that it radiates similar waveforms in all directions. However, if waveform dispersion does occur, and it is in a predictable pattern, the system might be able to compensate for it.

Antennae with larger-scale elements radiate low-frequency components, which results in a chirp-like, dispersive waveform. And, the waveform will vary at different azimuthal angles around the antenna. A multiband or OFDM approach is more tolerant of these dispersive antennae, but it still has its challenges. On the other hand, an antenna with small-scale elements, such as a planar elliptical dipole, is likely to radiate a compact, non-dispersive waveform, which is well suited to a commercial UWB application.

5.3.1 Antenna types

Antennae used for UWB can generally be classified as directional or omnidirectional. Directional antennae (also known as high-gain antennae) concentrate the electromagnetic energy in a smaller angle (as compared to an omnidirectional antenna). Antenna gain is measured in 'dBi', which is shorthand for dB relative to an ideal isotropic antenna. High-gain directional antennae can achieve 10 dBi or even 20 dBi gain, while an omnidirectional dipole antenna, for instance, typically demonstrates 2.2 dBi gain. In summary, a directional antenna offers high gain, a narrow field of view and a large antenna size. An omnidirectional antenna offers low gain, a wide field of view and small antenna size.

5.3.2 Antenna requirements

Matching networks are very common with narrowband antennae. These carefully engineered circuits navigate any impedance gaps between the RF section of the signal chain and the antenna; in other words convert the output impedance of the RF section to the appropriate impedance required by the antenna. The problem is that as the bandwidth grows, the complexity of the matching network becomes unwieldy. As a result, UWB antennae require a good impedance match between the antenna and the UWB radio. This must be designed in from the beginning because it cannot be fixed later with a matching network.

There remain other major challenges for designing UWB antennae. These include determining the performance versus directivity of the antenna, and selecting the optimal size for the application or level of performance. Most of these decisions are influenced by a great deal of antenna theory and an engineering-level understanding of antenna technology that are beyond the scope of this book.

5.3.3 Antenna availability

There is plenty of published theory on UWB antennae, dating back more than 30 years, and most is attributed to the military. There are however, few commercial UWB antennae available, although we can expect that many UWB antennae are in development and should be hitting the market soon. Depending on their requirements, UWB applications could use horn, reflector, dipoles or loop antennae. For some applications, UWB antennae might also be combined into arrays.

5.4 Radios built on cards vs. integrated designs

Regardless of which access technology is used, mobile applications are key drivers for integration. Specifically, mobile applications require small size and low cost. The challenge is that a radio doesn't only contain RF components, but it needs to transmit and receive data from a host processor to be useful. Inevitably, the radio also includes buffer memory and multiple I/Os, such as USB, PC interconnect (PCI), and other functions that increase the system's complexity. The objective is to include all of these functions into a single, low-cost component. The situation becomes more challenging because radios are made of both RF components and digital components, which do not integrate well using the same process technology.

For instance, the digital portion of the radio requires a large number of gates, so it benefits from small process geometry and is typically implemented in CMOS. The RF section, on the other hand, requires a well characterized process that is typically based on a larger geometry that is easier to implement in a manufacturing process such as SiGe.

Demands for lower costs are encouraging single-chip integration in CMOS. This is a technological challenge that is being addressed across the entire wireless industry. The end result will be a highly integrated radio that contains all of the components necessary for data transfer with digital interfaces such as USB or PCI on one side and an antenna interface on the other side.

The same integration evolution – from cards, to modules, to IC – has been followed by Bluetooth and WiFi devices. In both cases, products were initially made of multiple chips that were developed in different process technologies (CMOS for the digital part, SiGe for the RF part). As market demand and production volumes grew, so did requirements of lower costs and size. This resulted, in both cases, in a higher level of integration that led to a single-chip Bluetooth and a single-chip WiFi device. It is undisputed that UWB will follow the same evolution. The next challenge for the industry is how to integrate multiple radios that will be present in the same product.

The first example we'll consider is the PC. The picture in Figure 5.4 shows a PC motherboard reference design published by Intel. The south

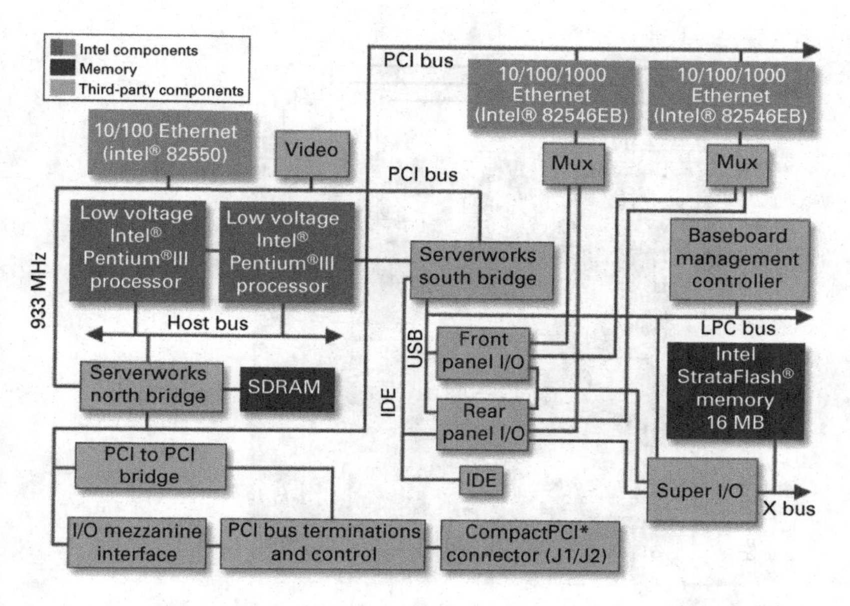

Figure 5.4 PC motherboard reference design

bridge chip manages I/O functions, e.g. for USB, audio or serial ports, the system basic input–output system (BIOS), the industry standard architecture (ISA) bus, the interrupt controller and the integrated drive electronics (IDE) channels. In other words, it contains all of the functions of a processor except memory, PC interconnect (PCI) and advanced graphics processing (AGP).

Note that radios for WiFi, Bluetooth, UWB and WiMAX are not part of the motherboard. One reason is that radios need certifications (such as from the FCC in the US and from other regulatory bodies in other countries) to be allowed to transmit. It is cost-prohibitive to send the entire motherboard through the various regulatory certifications. In the case of the PC, one way to achieve integration, reduce cost and shrink size is to combine multiple radios in a single chip. Various companies have already announced products and plans to integrate WiFi, Bluetooth and FM in a single 'wireless south bridge'.

The situation is different in the case of a handset, as shown in Figure 5.5. There are three main components in a handset: digital baseband, analogue baseband and RF transceiver. The RF transceiver interfaces to filters and the antenna. The analogue baseband interfaces

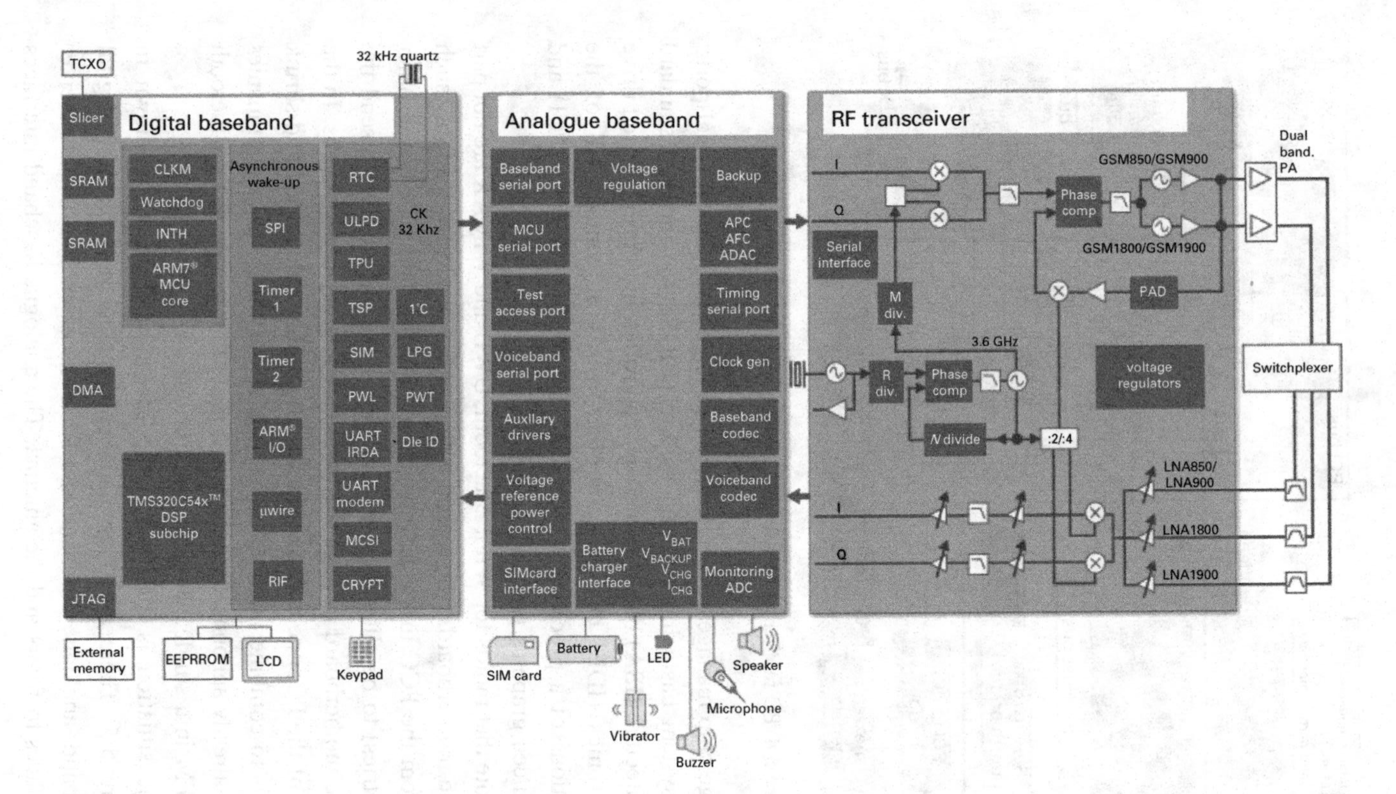

Figure 5.5 Handset reference design

to analogue I/Os, such as light-emitting diodes (LEDs), the microphone and speaker. The digital baseband interfaces to digital I/Os, such as a keypad, liquid-crystal display (LCD), external memory or other interfaces, such as secure digital input output (SDIO), USB or others (not shown in the figure).

Bluetooth, WiFi or UWB also interface to the digital baseband chip because they must carry data to the main host processor, similarly to the wired interfaces. There are two main paths to further integration: either combine these functions in the same chip, as in the PC example, or integrate the digital components of those functions in the digital baseband chip while leaving the RF components separated. The advantage of the first option is obvious: more functions in the same chip reduce the cost and size of the overall solution. The advantage of the second option is that all digital functions can follow the digital baseband process migration to a smaller geometry, while the RF components (which are much more difficult to migrate because they require process characterization and cannot take advantage of smaller geometries as much as their digital counterparts) can be combined together in a single chip.

So far, the industry has shown a preference for the first option, with initial integration of Bluetooth and WiFi. Some of the reasons for this are technical, and some are practical. For example, technologies for WiFi and Bluetooth might be developed by different groups within the same company. Or, the firmware requires a dedicated processor because of the real-time response requirements to run the protocols.

5.5 Summary

In the implementation of UWB into a larger system, there are a number of trade-offs which a designer needs to make along the way. Several of these trade-offs are discussed. Since UWB is a radio and may well be built into a system with multiple radios, there is always a possibility of interference between radios. As a rule, designers should always seek to create an environment where interference between radios is not possible, or if possible, is minimized through separation and coordination techniques.

On the question of integration, the objective is generally to select processes that reduce die size and cost, maintain performance and simultaneously use materials that make it possible to integrate all or part of the design into larger components. This is an optimization problem that tends to favour the use of CMOS wherever possible. When looking at packaging options for the resulting die, a number of trade-offs are possible, which can be used to minimize footprint, reduce cost and possess proper RF characteristics. In general, LTCC has an edge in the trade-off process.

In the area of antennae, there is a great deal of variability that will occur by application. While it is possible to use horn, loop or other antenna types, a great deal of the selection will depend upon the application requirements that govern directionality, size and cost. At the moment, there is a fairly thin offering of antennae to fit UWB requirements, but it is expected that this will increase as the market grows.

6 Upper-layer protocols

Douglas Adams once said, "Anything that was invented before you're born is normal and ordinary and is just part of the way the world works, anything that's invented between when you're 15 and 35 is technology, anything invented after you're 35 is against the natural order of things."

This chapter aims to explain the 'natural order' of some of the protocols that enable wireless connections. Consumers rarely see the technology or read the specification for a wireless protocol. Instead, they are in contact with the application layer that determines the normal and ordinary behaviour of the product and its high-level features. If you imagine a layered radio product, the top layer is the application layer. Underneath that, a standardized protocol determines how the radio interacts with the rest of the network. Examples of protocols include CW USB (or wireless USB), WiMedia Layer2 Protocol (WLP), Bluetooth, Wireless 1394 and ZigBee. Below the protocol is a common radio platform, which, depending on its capabilities, could simultaneously support multiple protocols. A diagram of this structure is included in Figure 3.1.

To help explain the various available protocols, consider a common application – sharing photographs. In this sample application, a consumer uses a camera phone to take pictures, and then wants to print the pictures on a local printer and send them to a website. The initial step is to establish a wireless connection between the camera and the printer. The first time this is done is difficult because the connection must be secure to avoid anyone intercepting the pictures. After that, when a trusted relationship between camera and printer is established, it should be easy. Regardless of the protocol used, consumers expect the process to be easy and secure. Depending on the protocol used, the degree of ease and security can vary, and product designers and planners need to determine which protocol offers the right balance of ease of use, performance, power, security and speed for the end application. For instance, Bluetooth is not as secure as CWUSB and WLP. And, although WLP can simultaneously connect to the printer and the PC, the same IP protocol that allows this flexibility is likely to add complexity to the ease of use of the product.

From the consumer's perspective, how the file transfers from camera to printer is immaterial, and the actual protocol doesn't play any role in the user experience. In fact, the user only becomes interested in the protocol when it stops working! For example, there could be a problem if the user moves the device out of range, the printer is already printing, a new device wants to connect at the same time or the camera accidentally disconnects.

All of these cases are treated differently by the various protocols, resulting in a different user experience. We will examine this particular example in more detail throughout the chapter in terms of each specific protocol. These are the types of case that a product designer or marketer needs to consider carefully before choosing a protocol. For each protocol, this chapter will cover:

- Main applications,
- System architecture,
- Main challenges,
- Protocol description,
- Strengths and weaknesses,
- Some more information on how the protocol solves the example in the introduction.

6.1 Certified wireless USB (CWUSB)

Anyone who has ever connected a device or peripheral to a computer is familiar with the USB port. There are already more than two billion wired USB connections in the world today, and, after its introduction, USB quickly rose to become the de facto external connection standard in the PC industry.

6.1.1 Main applications of CWUSB

Designed as a cable-replacement technology, CWUSB is primarily targeted at PCs, peripherals and mobile devices. The intent is to offer high-speed wireless connections to nearby equipment.

Key home uses include: digital video camcorder, portable MP3, printer, scanner, external hard drive, PDA, tablet PC, wireless speakers, stereos, HDTVs, video recorders and PCs (see Figure 6.1). Target enterprise applications include data backup, printer connectivity, scanner connectivity and synchronization of mobile devices (such as a PDA with a cellphone or notebook computer).

6.1.2 System architecture

The CWUSB is designed as a high-speed host-device connection, where a host (PC) controls the link to its peripheral devices. Using a point-

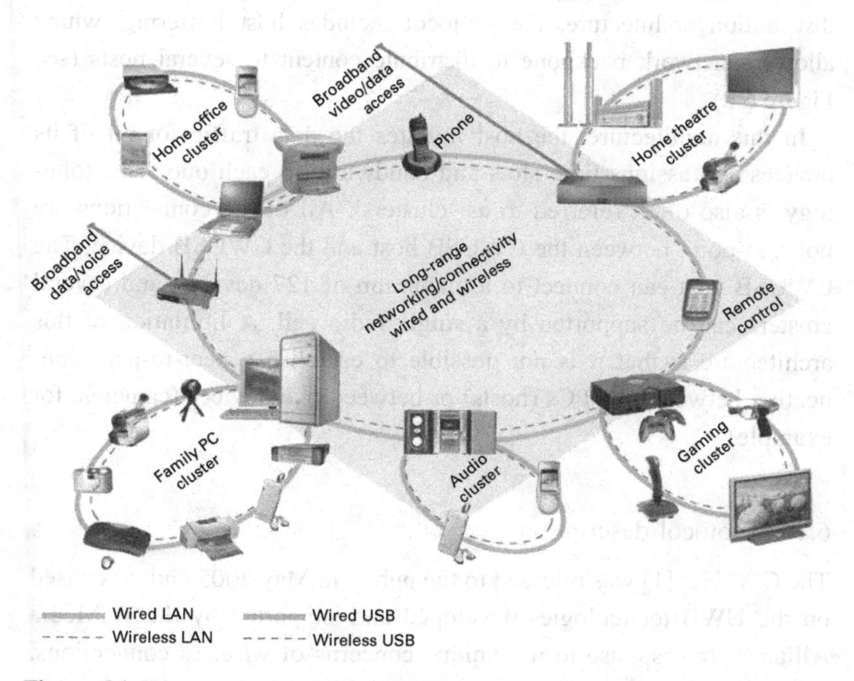

Figure 6.1 Home-use scenarios for WUSB

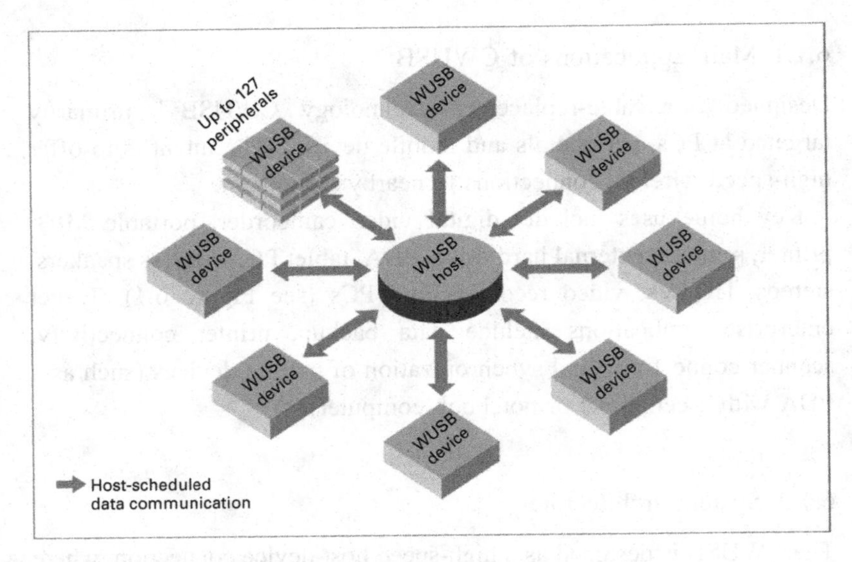

Figure 6.2 WUSB host communications

distribution architecture, the protocol includes host buffering, which allows a network backbone to distribute content to several hosts (see Figure 6.2).

In this architecture, the host initiates the data traffic for all of its devices and assigns time slots and bandwidth to each one. This topology is also often referred to as 'clusters'. All of the connections are point-to-point between the CWUSB host and the CWUSB device. The CWUSB host can connect to a maximum of 127 devices, and several clusters can be supported by a single radio cell. A limitation of this architecture is that it is not possible to establish a peer-to-peer connection between two PCs (hosts) or between two devices (cameras, for example).

6.1.3 Protocol description

The CWUSB [1] was released to the public in May 2005 and was based on the UWB technologies developed and supported by the WiMedia Alliance. In response to the unique concerns of wireless connections, CWUSB has built-in features to handle noise, fading and conversion delays. The USB-IF reports that the protocol demonstrates 75% efficiency in data transfers over a typical 480 Mb/s link.

6.1.4 Strengths and weaknesses

With a projected bandwidth of up to 480 Mb/s, CWUSB can support several high-bandwidth data streams simultaneously. The actual application-layer throughput will be somewhat lower, given the various overheads provided by protocol, PC drivers, etc. In general, it offers a simple, low-cost implementation and supports up to 127 devices per host. Backward compatible with standard USB, CWUSB reportedly offers the same level of security as its wired counterpart.

Given that CWUSB is a whole new protocol that contains many of the original USB concepts but is optimized for wireless, it requires specific CWUSB drivers to operate. That means that a CWUSB dongle can't just be plugged into an existing USB port on the PC and be expected to work; it requires the user to install a new PC driver.

6.1.5 Main challenges

Users converting from standard USB will come to CWUSB with high-performance expectations. The main challenges for CWUSB are to maintain high performance without drawing too much power, enabling an appropriate level of security and ensuring ease of use.

The main usage scenario for CWUSB involves file transfers. The performance comparison is to the wired USB counterpart. Different products offer different throughputs: the highest-performance hard drives can support up to 250 Mb/s, but most current products are limited by other factors, such as central processing unit (CPU) performance, memory access speed or power-consumption considerations, and it is typical to see 50–100 Mb/s performances in wired USB. CWUSB needs to be comparable in performance, at a range of 0–30 ft.

Power management is a concern for all wireless technologies. The CWUSB protocol addresses this issue by incorporating multiple activity modes for the radio, including sleep, wake, listen and conserve.

Because it is targeted at residential and enterprise applications, across distances up to 30 feet, CWUSB was designed with built-in security protocols and authentication procedures, which were designed not only to provide a private connection against casual eavesdroppers, but also to be resistant to any malicious attacks. It also encrypts data during transmission to heighten the security of the link.

One of the key benefits of wireless connectivity is convenience, so maximizing ease of use is of paramount importance for all protocol developers. In fact, the CWUSB specification provides an easy way for consumers to connect CWUSB devices and hosts. The relatively long-distance links of CWUSB require the host and device to maintain a connection in an environment that might be crowded with interference. By operating between 3 and 10 GHz and using the UWB radio, CWUSB avoids interference from incumbent 802.11 WLAN systems, which operate at 2.4 GHz.

6.1.6 Application example

For the camera phone–printer–web connection mentioned in the introduction, CWUSB would work this way: the consumer pushes a button on the printer and one on the phone. Each then displays a series of numbers. Then, the user compares the two numbers on the phone and the printer. If they match, the user simply presses a button on the phone to accept the connection. At that point, the photograph can be transferred to the printer. To transfer the same picture to the web, the user needs to follow the same process with the PC.

Since CWUSB mimics the USB cable, the camera can only be connected to one host at a time, much like connecting and disconnecting a USB cable. It is possible for all CWUSB units to store previous connection information locally; so that they can remember all the devices they connect to, and reconnect more simply.

6.2 WiMedia layer-two protocol (WLP)

The WLP was developed by the WiMedia Alliance to enable internet protocols (IP) over WiMedia UWB radios. Even though IP is also supported by WiFi, WLP does not compete with WiFi; it is designed to complement it. WLP and WiFi are very different – both for reasons driven by marketing requirements and the nature of the underlying radio.

For instance, WiFi is primarily used to provide Internet access to laptops, while WLP is envisioned as supporting ad-hoc personal area networks (PANs). It doesn't bring Internet access to laptops, but rather utilizes the laptops to bring Internet access to other devices, as described later. Ad-hoc networking is where devices come together and assign themselves 'ad-hoc' network addresses rather than rely on a managed server to keep track of Internet addresses. To date, there has been very little ad-hoc networking.

While the main emphasis of WLP is on devices, PANs and ad-hoc networking, nothing precludes Internet connectivity. In fact, as broadband metropolitan wireless Internet access evolves with systems such as WiMAX, the phone or laptop will become the Internet access point for the PAN. In these cases, net access will no longer be tied to 'hot spots' and local servers, so ad-hoc networking will be able to access the Internet.

6.2.1 Main applications

The WLP is targeted at high-bandwidth applications where ad-hoc wireless networking can form IP access points to create connections in residential and enterprise settings. Unlike CWUSB, WLP is a protocol with a developing market. As more and more devices transition to IP and people begin to understand that they can connect wirelessly at Internet access points, the applications for WLP will open up.

The first applications will probably be peer-to-peer network connectivity among battery-powered devices, such as cameras, MP3 players and mobile phones. This protocol allows, for example, a user to share a song on an MP3 player by transferring directly to a friend's MP3 player without needing a PC.

In the past, many engineers have resisted using IP for peer-to-peer connections because the protocol overhead has been viewed as too expensive. As processors and memory get cheaper and cheaper, software becomes a larger component of overall system cost. Today, applications that use software written for the Internet are less expensive to implement than custom applications. So, as the trend to using IP in device-to-device communications grows, so will the applications for WLP.

6.2.2 System architecture

The WLP can be used to support battery-powered devices that require peer-to-peer connectivity in a personal area network (PAN). A PAN, for example, could consist of an ultra-mobile PC (UMPC), a phone and a digital camera.

In a WLP system, devices communicate directly with other neighbouring devices, and they can also communicate with other client devices and nodes (such as Ethernet stations) by using the WLP client bridge function (see Figure 6.3).

These WLP bridges are WLP nodes that are capable of bridging to other IEEE 802.3 networks, such as the Ethernet. The design goal was to place the burden of any complexity on the bridge so that the WLP

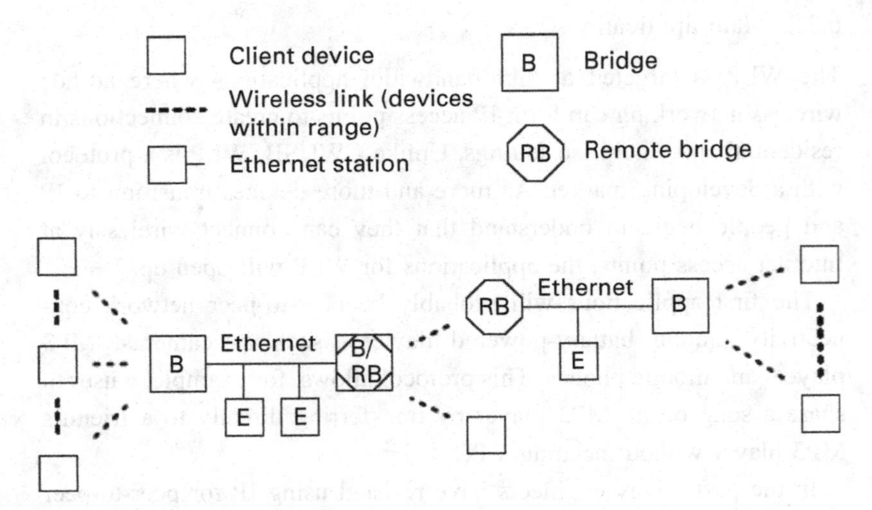

Figure 6.3 WLP client bridge function

nodes would remain simple. For example, a WLP bridge could plug into an Ethernet port and allow a WLP link to corporate networking. This could provide secure office or cube-wide networking for an individual's portable devices. Since WLP bridging is a pure level-2 IEEE 802.3 bridge, WLP devices would be indistinguishable from devices directly connected to the corporate network.

The WLP bridges are particularly useful for PANs. Any device, such as a laptop, UMPC or mobile phone, can act as an Internet router if it has access to the Internet through some sort of wireless service. If this were the case, other nodes in that WLP security set (WSS) would have access to Internet connectivity.

The WLP does not currently support mesh networking. (Mesh networking is when one node forwards frames to another node that the first node cannot hear.) In WLP, bridges can perform this function to a limited extent. The designers of WLP felt that it was better initially to deploy WLP without mesh networking and extend it later if or when real needs should arise.

6.2.3 Protocol description

The WLP is a protocol adaptation layer (PAL) that builds on the WiMedia UWB common radio platform to enable TCP/IP services. It uses a WiMedia-defined protocol ID value to identify WLP frames. The protocol defines four types of frame: standard data frames, abbreviated data frames, control frames and association frames.

The WLP supports Internet Protocol version 4 and Internet Protocol version 6 for data networking using the WiMedia MAC, transporting network-layer packets over the WiMedia radio platform.

Security is addressed through the concept of a WLP service set (WSS), which is a named set of devices with a security relationship. The WLP exploits the security mechanism of the WiMedia MAC; however this mechanism requires a shared secret, to generate encryption keys. The name of a WSS is public and can be seen by anybody within range, but the cryptographic key necessary to see traffic is held only by members of the WSS.

To allow new members into a WSS, some sort of authentication scheme was necessary. The WLP specifies a set of message exchanges that support authentication. The authentication scheme is based on the 'simple config' protocols defined by the WiFi Alliance, but it has been extended to include the 'numeric compare authentication method' that is used by Certified Wireless USB (CWUSB). Since many WiMedia equipped devices will support both CWUSB and WLP, the numeric compare method provides a consistent user interface for first-time association of devices.

Even when a WLP node is acting as a bridge, it does not determine how other nodes access the medium. The function of bridging is completely separated from medium access, as WLP nodes are free to use either the distributed reservation protocol (DRP) or prioritized contention access (PCA) to determine when to transmit and listen (see Chapter 3).

Not all nodes in a WSS need to be in range of one another to share access. The WiMedia MAC accounts for the fact that all nodes in a beacon group may not be able to communicate with one another.

6.2.4 Main challenges

Power management and quality of service (QoS) were two key challenges that WLP needed to address. The WLP was designed to assume that all devices in the PAN would be battery powered. In host-device architecture, such as CWUSB, the host is in charge of power management. It either tells devices to sleep or grants permission to sleep. In contrast, WLP provides a structure where peer devices can share the power load and distribute it equitably. Each device is autonomous, but it shares in the burden of keeping the network alive. The WLP takes advantage of the power-saving features built into the WiMedia MAC in order to distribute power consumption among peer devices.

All active devices are required to beacon – or announce their presence – in a beacon group. Sleeping devices can miss the beacon period for several seconds before they are considered disconnected from the WSS. Of course, all of the devices cannot sleep simultaneously. At least one node has to continue to beacon so that all of the clocks can remain synchronized. This node is called the anchor. The WLP devices advertize in their beacon whether they have limited power on not. If a mains-powered device is available, it will become the anchor. If not, the responsibility to be the anchor will be shared among peers and transferred every few minutes to another node.

Because WLP offers raw bandwidth peaking at 480 Mb/s and real-world scenarios dictate that most high-bandwidth transfers are point-to-point, many engineers believe that QoS is a non-issue for WLP. In general, for proper QoS, streaming media applications need to have priority access so that they can operate at high quality even in the presence of large file downloads.

The PCA mechanism built into the WiMedia MAC allows for applications that need low-latency streaming to co-exist with high-bandwidth file transfers. In real-world scenarios, this means that you could use WLP to stream audio to a headset while files are being downloaded and using all available bandwidth.

6.2.5 Strengths and weaknesses

The WLP's main strengths are that it is designed to work with the WiMedia UWB common radio platform and it takes advantage of IP networking, which is making great inroads into the networking market. Its primary weakness is the fact that it targets an emerging market, and is dependent on IP-network node deployment for its success.

6.2.6 Application example

For the camera phone–printer–web connection mentioned in the introduction, devices equipped with WLP would connect with each other similarly to the CWUSB case. It is likely that the initial connection will take longer than CWUSB because setting up the IP network is much more complex than establishing a host–device connection with CWUSB.

The main difference is that the camera can maintain the connection with both the PC and the printer at the same time. That means that one can start transferring pictures to the web through the PC while printing.

6.3 Bluetooth

First released in 1998, Bluetooth is the protocol covered in this chapter that is probably best known by consumers. This is in large part due to excellent marketing strategies as well as early market penetration in wireless headsets for mobile phones – which had the good fortune to occur simultaneously with government legislation that required hands-free usage of phones in automobiles. The success of Bluetooth technology is notable. In November 2006, the billionth Bluetooth device was shipped. [2]

Geared towards voice and data applications, Bluetooth supports three different range scenarios, depending on the device class:

- Class 3 radios – range of up to 1 metre (3 feet),
- Class 2 radios – range of up to 10 metres (30 feet), and commonly found in mobile devices,
- Class 1 radios – range of up to 100 metres (300 feet), and used primarily in industrial applications.

The Bluetooth specification is developed and managed by the Bluetooth Special-Interest Group (SIG), which consists of more than 7000 member companies in the fields of telecommunications, computing, automotive, music, apparel, industrial automation and networking. Its standard data rate for version 1.2 is 1 Mb/s, and for version 2.0 it is up to 3 Mb/s. The most commonly used Bluetooth radio is Class 2, which transmits up to 2.5 mW of power. Radios are usually powered down when not in use and this allows a very long battery lifetime for most Bluetooth devices.

Recently announced, Bluetooth 3.0 is the new Bluetooth wireless standard currently being developed by the Bluetooth SIG and WiMedia Alliance. This new high-speed Bluetooth standard is expected to offer transfer rates of 400 Mb/s at close range and 100 Mb/s at 10 metres, vastly extending the application scenarios for Bluetooth by operating in the unlicensed spectrum about 6 GHz. It is expected that the current radios operating at 2.4 GHz will be used to establish the connections, while the UWB radio will be used as a high-speed pipe for the high throughput transfer. In these cases, UWB technology will be compatible with Bluetooth radios. The Bluetooth technology will maintain its core attributes. Products equipped with Bluetooth 3.0 are expected to be available in 2008.

6.3.1 Main applications

Bluetooth's main application is as a cable replacement. For instance, it can be used to connect office peripherals, wireless headsets, keyboards, mice, PDA synchronizations and transfers, mobile computing and mobile phones. Bluetooth will enable additional applications, such as streaming media, wireless presentations and transferring still photos at rapid rates. It is widely expected that Bluetooth will be the most-used protocol running over UWB for handset applications, because of the existing installed base.

6.3.2 System architecture

Currently released Bluetooth products operate in the unlicensed ISM band at 2.4 GHz and use a Gaussian frequency-shift keying (GFSK) modulation scheme, which applies Gaussian filtering to the modulated baseband signal before it is applied to the carrier. Bluetooth 3.0 products will operate in the 6.0–9.0 GHz band.

Bluetooth uses a spread-spectrum, frequency-hopping, full-duplex signal at a typical rate of 1600 hops/s. The frequency-hopping transceiver helps to combat interference and fading. For basic rate Bluetooth, the specified symbol rate is 1 megasymbol per second (Msps) supporting the bit rate of 1 Mb/s. For enhanced data-rate Bluetooth, the gross air bit rate is 2–3 Mb/s. For Bluetooth, data rates are expected to improve dramatically – up to 100 Mb/s.

The devices within a Bluetooth piconet use a specific adaptive frequency-hopping pattern that is automatically determined using certain fields in the Bluetooth specification and the clock of the master device. The basic adaptive hopping pattern deploys pseudo-random ordering of the 79 frequencies in the ISM band to avoid interference and improve co-existence with other wireless devices. Data is transmitted between Bluetooth-enabled devices using a time-division duplex (TDD) scheme, where data are transmitted in packets that are positioned in time slots.

During typical operation, a single radio channel is shared by a group of Bluetooth devices with a common clock and frequency-hopping pattern. The basic system architecture for Bluetooth is a master–slave approach. In this scheme, one device provides the synchronization reference and is known as the 'master'. All of the other devices are known as slaves. When a group of devices is synchronized in this fashion, the PAN is referred to as a piconet. Bluetooth's system architecture can support up to seven slaves per master.

In Bluetooth [3] a WiMedia PHY and MAC system is added under L2CAP (see Figure 6.4). This allows the device to use the Bluetooth 1.2 system for device discovery, authentication and to maintain the connection. Once a service is requested that requires a file transfer or a high-speed streaming application, the system switches to UWB to complete the data communication. This provides users with the best of both worlds: the proven interoperability, ease of use and long battery lifetime provided by Bluetooth, and the high throughput provided by UWB.

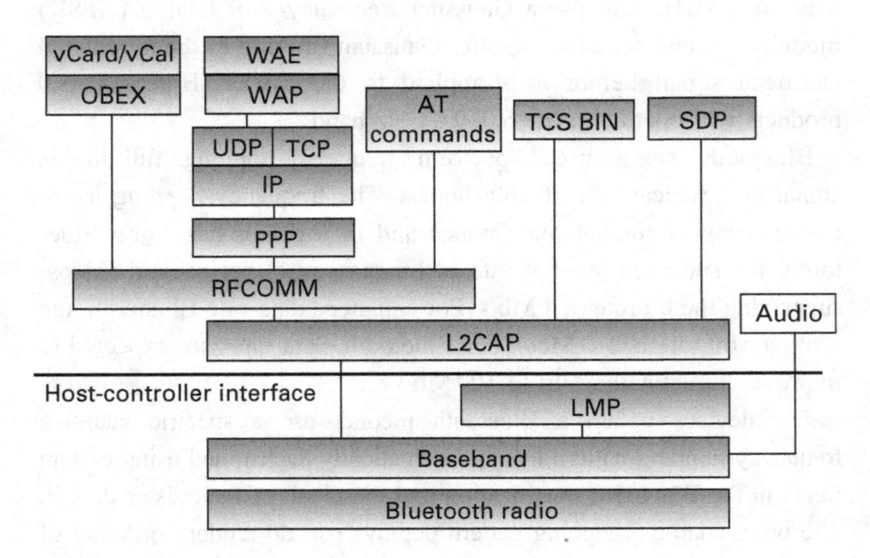

Figure 6.4 Bluetooth protocol stack

6.3.3 Protocol description

The ultimate objective of the Bluetooth protocol is to allow applications to interoperate with each other. To do this, they must run over identical protocol stacks. In an effort to ease implementation, the Bluetooth specification is divided into the following categories:

- Core specification: describes the protocol stack up through the L2CAP layer and the characteristics of each of the relevant protocols, as well as the relationship between them.
- Profile specifications: define a set of features required to support a particular usage model or set of usage models. (A profile specification document describes how to use the protocol stack to implement a given profile. Each of these profiles, however, still uses a common Bluetooth data link and physical layer.)
- Transport specifications: define the physical interfaces that can be used to implement the host–controller interface (HCI). (The HCI transports are used in products that choose to separate the implementation of the host and controller functions.)

Figure 6.4 shows the complete Bluetooth protocol stack as identified in the specification. Interoperable applications supporting the Bluetooth usage models are built on top of this stack.

The WiMedia UWB system is inserted below the HCI and controlled by L2CAP. This allows for reuse of most of the software and the learning of many years of interoperability improvement of Bluetooth 1.2.

6.3.4 Main challenges

For versions up to 2.0, one of the major challenges for Bluetooth is that of mitigating interference. Adaptive frequency-hopping techniques were added to compensate for this problem. Concerns over interference are minimized in Bluetooth 3.0 because the radio will operate above 6 GHz, well out of the crowded 2.4 GHz range used by wireless network and other portable devices. Another concern for Bluetooth is security, and this is an issue that the Bluetooth SIG continues to address.

All of the devices in a Bluetooth piconet have a unique 48-bit identity number. The first device identified in the piconet becomes the master, and it sets the 1600 frequencies that will be used each second across the band. All of the other devices in the network synchronize to the master. The OEMs can select among three modes of security for Bluetooth access in their devices:

- Security mode 1: non-secure,
- Security mode 2: service-level enforced security,
- Security mode 3: link-level enforced security.

Devices and services are known either as 'trusted device' or 'untrusted device'. A trusted device, which has been paired with another device in the piconet, has access to all services. Services can have one of three security levels:

- Services that require authorization and authentication,
- Services that require authentication only,
- Services that are open to all devices.

There has been much talk in the popular and technical media about lapses in Bluetooth security, coining terms such as 'bluejacking', and 'bluebugging'. Many of these cases involved mobile phones, and the Bluetooth SIG discovered that most of these cases involved incorrect implementation of the Bluetooth security features. So, while product planners and designers can feel fairly confident about Bluetooth security, they must ensure that it has been properly implemented.

6.3.5 Strengths and weaknesses

Each version of Bluetooth has its different strengths, whether handling voice or data transmissions. Bluetooth 2.0, for instance, is particularly well suited for wireless mobile headsets because of its low-cost low-power, and low data rates. With Bluetooth 3.0, manufacturers will have access to high-speed connections as well as the low-power, low-cost advantages historically associated with Bluetooth. In addition, operations above 6 GHz will ensure the best co-existence with other wireless products (Bluetooth will be backward compatible with earlier versions of the protocol that operate at 2.4 GHz).

6.3.6 Application example

For the camera phone–printer–web connection mentioned in the introduction, Bluetooth would work this way: the user connects the camera to the printer as with the current version of Bluetooth. After the connection is established, the devices automatically switch to UWB and establish a fast transfer to the printer. After that, the same process is followed with the PC to transfer the pictures and send them to the web.

6.4 Wireless 1394

Much like CWUSB, Wireless 1394 is based on a proven high-speed wired connection specification already in existence: in this case, IEEE 1394, which began as technology from Apple Computer known as 'FireWire'. In May 2004, the 1394 Trade Association approved a new protocol adaptation layer (PAL) for IEEE 1394 over IEEE 802.15.3, which enabled 'Wireless FireWire' product development.

6.4.1 Main applications

The main applications for Wireless 1394 include high-speed data transfer, especially between set-top boxes (STBs), HDTVs, tuners and DVD players. Wireless 1394 should provide an edge for applications that require good QoS, such as video streaming and audio–video synchronization.

6.4.2 System architecture

Wireless 1394 implements a peer-to-peer or 'daisy-chain' network topology, meaning that all of the devices connected in the network communicate with each other. The system can support as many as 63 devices on a single connection.

6.4.3 Protocol description

Developed by the Wireless Working Group of the 1394 Trade Association, the Wireless 1394 protocol is designed as a standard convergence layer between WiMedia or other MACs (such as IEEE 802.15.3) and applications that were previously developed for wired 1394.

The WiMedia Alliance's MAC Convergence Architecture (WiMCA) serves as the platform for the wireless 1394 protocol adaptation layer (PAL). The PAL enables the use of IEEE 1394 devices and protocols in a wireless environment. Data speeds up to 480 Mb/s are supported, and compatibility is maintained with existing wired 1394 devices.

Wireless 1394 builds upon the 1394 infrastructure, including the same data formats, connection-management schemes and time-synchronization procedures.

6.4.4 Main challenges

The main challenge for Wireless 1394 is marketing, not technical, as it is in direct competition with Wireless USB, which can leverage the market share of wired USB. Industry has reduced the support to both wired and wireless 1394 in recent years.

6.4.5 Strengths and weaknesses

The strengths of Wireless 1394 are the same as those for the wired IEEE 1394, including excellent quality of service (QoS) enabled through the wireless MAC, low implementation complexity and the ability to install several devices on a single bus.

6.4.6 Application example

Wireless 1394 has not found a lot of momentum in the market. If it does, products are likely to utilize it for video transfer, where QoS is more useful. A camera phone–printer–web connection application would behave more similarly to WLP than to the other protocols described in this chapter. For example, with Wireless 1394 the camera can maintain the connection with both PC and printer at the same time. That means that one can start transferring pictures to the web through the PC while printing.

6.5 Association

Assume for a moment that there are five or six devices within hearing range of each other, which all support a common protocol stack. One of the first challenges in establishing communication between any two devices is to identify one another and to exchange cryptographic keys so that a securely encrypted link may be established. This process is known as association and resides at the bottom edge of a protocol stack, as indicated in Figure 6.5.

Each protocol stack has the potential of establishing a different method of associating devices. However, since the options for creating a secure link are somewhat limited and since there is benefit to the technical community in using common techniques (as opposed to retraining consumers), most special-interest groups attempt to use common methods unless their specific applications require them to differentiate.

There are three techniques that are used on UWB devices: cable, PIN (personal information number) and NFC (near-field communications). The first technique, cable association, allows for two devices to be connected to each other initially through a cable. A common, cryptographic secret is shared between the devices, and is used to encode further

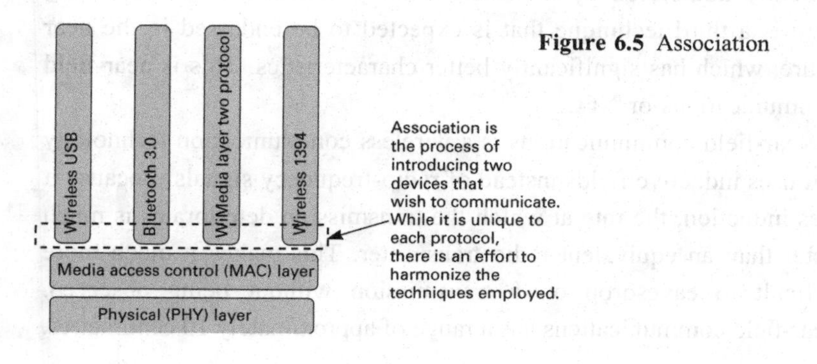

Figure 6.5 Association

communications. After the secret is exchanged, the cable is removed and the wireless link can be used from this point forward. Obviously, this is a somewhat sub-optimal technique as consumers see it as requiring a wire to become wireless. From a product standpoint, irony is rarely a strong selling point. However, from a security standpoint, this approach is extremely robust. In the case of WUSB, this approach makes a lot of sense. Many applications use the WUSB cable to provide power to the peripheral device. In a more mobile context, the wire may still be present to a docking cradle. Using this path as a method to establish association may well prove to be quite convenient for this class of consumers.

The second technique involves a PIN. Imagine a situation in which two devices are present. Call them A and B in a flash of naming brilliance. Each device has a four digit PIN imprinted on a sticker which is permanently affixed to the unit. A button is pressed on device A to transmit its PIN. Device B has a numeric display, which shows a four digit PIN. If the user sees the number that is printed on A's sticker, the user will press a button on B to accept the link. If a number other than the one that is printed on A's sticker appears, a third device is present and is attempting to insert himself into the middle of the exchange. This is known as a 'man in the middle' attack. The link should not be accepted. This approach has an assumption that at least one of the devices involved has a numeric display. This is again a very secure approach, but not necessarily a very elegant one. In some instances, the display may add cost to price-sensitive applications or may disturb the aesthetics of a device.

At the time of this writing, these are the two approaches which are presently authorized by WiMedia and by Wireless USB. However, there is a third technique that is expected to be endorsed in the near future, which has significantly better characteristics. This is near-field communications or NFC.

Near-field communications is a wireless communication technology that uses inductive fields instead of radio-frequency signals. Because it uses induction, the rate at which the transmission deteriorates is much faster than an equivalent radio transmitter. This makes it much more difficult to eavesdrop on a transmission without being observed. Near-field communications has a range of approximately 10 centimetres.

Say, for instance, that a mobile phone is attempting to use a large panel display to show photos. Both of the devices would have an NFC logo strategically placed above the NFC transciever. The consumer would touch the logo on the mobile phone to the logo on the display. While physical touch is not actually required, this act has the effect of bringing the two devices within communication range. The cryptographic secret would then be exchanged over the NFC link. This information would then be sufficient to establish a secure link over the UWB radios. The NFC communication link can then be dropped. To the consumer, this type of engagement is very intuitive. Touch anything with which you wish to communicate.

While the NFC approach has been known for some time now in the PAN community, it was not mature enough to incorporate into practice. The standards were not yet finished. The number of chip suppliers was very limited. Because of this, it was necessary to delay introduction of that authentication technique.

6.6 Summary

Ultra-wideband standards have been developed recently that address consumer-market applications and work with or directly support the protocols that determine how the radio interacts with the rest of the network. The protocols examined in this chapter include USB-IF (or wireless USB), WLP, Bluetooth, Wireless 1394 and ZigBee. Each of these protocols is a common radio platform, which, depending on its capabilities, could simultaneously support multiple protocols.

Certified WUSB was designed as a cable replacement for consumer electronics, and was primarily targeted at PCs, peripherals and mobile devices. The aim was to offer high-speed wireless connections to nearby equipment.

The WLP was developed by the WiMedia Alliance to enable internet protocols (IP) over WiMedia UWB radios. It is envisioned as supporting ad-hoc personal area networks (PANs). It doesn't bring Internet access to laptops like WiFi, but rather utilizes the laptops to bring Internet access to other devices.

Newly announced Bluetooth 3.0 will enable additional applications such as streaming media, wireless presentations and transferring still photos at rapid rates. It is widely expected that Bluetooth will be the most used protocol running over UWB for handset applications, because of the existing installed base.

Though Wireless 1394 is facing a challenging marketing battle with Wireless USB, 1394 should provide an edge for applications that require good QoS, such as video streaming and audio–video synchronization. The main application for Wireless 1394 is high-speed data transfer, especially between set-top boxes (STBs), HDTVs, tuners and DVD players.

In each of the supported protocol stacks in the UWB ecosystem, there is a need to designate which two devices are attempting to communicate. To maintain a secure communication, these two devices need to exchange a cryptographic key, which is not known to others. The process is known as association. At present, there are three techniques either deployed or planned. The first of these is the cable-association model wherein a cable connects two devices (for the first introduction only) to enable the cryptographic keys to be exchanged and then the UWB link is activated. The second technique is a PIN approach where one of the devices is able to display a PIN, which is recorded on the second device in a permanent manner. And finally, the third approach to association involves the use of an NFC link, which is extremely short range and on which eavesdropping is very difficult. In this situation, the consumer touches two devices together to indicate his desire to have them communicate. The cryptographic key is exchanged across the NFC link and the UWB link is activated immediately afterwards.

References

[1] www.usb.org/developers/wusb/, accessed 5 October 2007.

[2] www.bluetooth.com/Bluetooth/Press/SIG/BLUETOOTH_WIRELESS_TECHNOLOGY_SURPASSES_ONE_BILLION_DEVICES.htm, accessed 4 October 2007.

[3] https://programs.bluetooth.org/apps/content/_doc_id=44515, accessed 5 October 2007.

7 Ultra-wideband standardization

To start this discussion on standards, one should understand that UWB standards and specifications will not be generated by a single standards organization. There are a number of organizations involved in the effort and each of these is engaged for a specific purpose. The following overview gives a flavour of how the division of labour is structured between Ecma International (Ecma), the International Standards Organization (ISO) and the European Standards and Technical Institute (ETSI) in the development of UWB.

7.1 Ecma International

The standards organization leading the effort in the development of the UWB physical layer (PHY) and the media access control layer (MAC) is Ecma International. Ecma was initially focused on developing standards for the European computer markets when it was created in 1961, [1] but has since expanded its charter to cover standards in software, consumer electronics and communications on an international scale. Ecma is responsible for developing such well known standards as the DVD (digital video disc) and emerging standards such as near-field communications (NFC), which uses inductive coupling to establish a link between smart cards (credit cards) and their readers. As a side note, the NFC techniques are likely to emerge as a means of association in next-generation wireless LAN and PAN devices.

In UWB, the two principal standards that Ecma is developing are ECMA 368 [2] and ECMA 369. [3] ECMA 368 includes both the PHY and MAC definition while ECMA 369 contains the interface definition between the PHY and MAC. The first version of these two standards was written based upon the regulatory conditions present in the United States. At the time that ECMA 368 and 369 were written, the US regulations were the only ones which were sufficiently stable to use. Later versions will expand the standard to describe variations needed for other regulatory environments.

Many of the companies responsible for driving the standards in Ecma are also present in the other bodies that are relevant to UWB. Committees in ETSI, WiMedia Alliance, the Bluetooth Special Interest Group and the Universal Serial Bus Implementers Forum are populated by many of the same companies who are working together globally to enable UWB. (These special interest groups will be discussed in greater detail in Chapter 9.) By having a common group of major companies working across all of these organizations, it is possible to coordinate work in a variety of different standards and industry fora.

Standardization work was initiated by Ecma member companies when it became clear that the initial effort which was conducted in IEEE 802.15.3a was fatally stuck. The IEEE committee members were unable to reach a conclusion on the modulation techniques that would be used in the standard. The IEEE committee voted to discontinue work. Shortly thereafter, Ecma members moved the work to Ecma technical committees and were able to complete a standard.

7.2 International Standards Organization (ISO)

Once Ecma completed work on 368 and 369, the next step in the process was to advance these standards into ISO for approval there. The Ecma standards were presented to ISO for consideration as an international standard. Whereas most standards groups are regional, as in the case of ETSI, or industry based, as in the case of the Institute of Electrical and Electronic Engineers (IEEE), ISO is populated by national administrations.

Approval of a standard by ISO effectively means that individual administrations have recognized the standard for the purposes of international law. Specifically, a network of trade agreements exist, which is focused at eliminating trade barriers around the world. By obtaining ISO recognition, a standard is protected by trade-treaty provisions. This will be discussed in more detail later in this chapter.

In addition to trade-treaty protection, ISO-standard approval also makes it easier for standards to be referenced by countries and other standards organizations. For instance, if an administrative body such as the European Commission wishes to encourage technical development within their region, they may instruct a regional standards body, in this case ETSI, to develop a given technology such as UWB. If an existing ISO-recognized international standard exists, it can be referenced for local use instead of developing a new standard from scratch. The ECMA standards which have been approved by ISO are ISO/IEC 26907 and 26908.

7.3 ETSI

The third major standards organization presently involved in UWB is ETSI. In Europe, ETSI is a quasi-governmental body. Its membership comprises both government members as well as industry representatives.

As part of the mandates issued by the European Commission (EC), which will be discussed in Chapter 11, ETSI was tasked to develop standards necessary to enable the introduction of UWB in Europe. In this case, standards could have been interpreted to mean the definition of the radio, definition of the higher-layer protocols or definition of the regulatory tests needed to establish compliance of a radio with regulatory requirements. However, ETSI chose to interpret the mandate as requiring development of testing standards.

In the international-standards community, it is normal practice (but not a hard obligation) to avoid having a single standard effort under development in two or more competing standards bodies. When ETSI received the mandate from the EC, there was work already under way in the IEEE (which has since ceased operations and is no longer relevant) on the UWB radio standards. The only remaining piece of standardization that was required was the development of measurement standards to establish compliance to European regulatory requirements. Because of this, the scope of ETSI's standardization effort is limited with regard to UWB. The development of these standards is dependent upon the completion of European regulations and so has only recently been taken up in earnest when the European allocations for UWB were announced in March 2007.

In addition to its formal obligations under the EC mandate on UWB, ETSI is also tasked with facilitating the successful launch of technologies that are deemed to be important to Europe. They do this by working with special-interest groups and other bodies to help define how compliance and interoperability testing is performed. ETSI's extensive expertise in these forms of testing is offered on a consulting basis to groups who are more intimately involved in the development of the technology. In this way, it is possible to avoid some of the failures that may otherwise occur to delay the emergence of a technology.

In UWB, a relationship has been established between ETSI and the WiMedia Alliance. In this relationship, ETSI is consulting with WiMedia to help automate testing processes and to improve both the repeatability and accuracy of testing being done.

This is an overview of the principal players in UWB standardization. For the most part, companies who produce end products that happen to incorporate a UWB radio chip at a system level will have little need to participate in these bodies. However, if your product is a UWB radio chipset, you may wish seriously to consider participation in one or more of these bodies to ensure that future development considers the needs of your customers.

In this next section, there is a discussion of the standards strategy that is being employed with regard to UWB. But, before getting into that discussion, it is useful to understand a little more about the standards environment on an international level.

7.4 An international perspective on standardization

Each national administration has a different view of the role of standards in developing its economy and the government's participation in standardization. Some countries believe that the government should take a very active role in the development of industry while others take a much more hands-off approach. The standards development bodies that evolve are frequently a reflection of this variable.

In the USA, government and standards are two very separate things, which interact minimally. Industry organizations enact standards or specifications, which individual companies either embrace or reject based upon their commercial interests. The government normally does little to nothing to compel industry to use standards. Companies support standards if they believe that they can profit from doing so and use proprietary techniques when this approach appears more effective. In this policy, it is possible to see a reflection of the USA's tendency to minimize government involvement in business. There is a tendency for US-based engineers to think of this as the only model for standardization and to skew their thinking based upon this assumption. The US model is substantially different from those used in other parts of the world.

In Europe, ETSI is a quasi-governmental body. It is a normal behaviour for the European Commission (EC) to issue mandates that task ETSI to create standards on a specific topic. In the case of UWB, the European Commission issued its first mandate, which included a directive to ETSI to undertake standardization of UWB. This level of government involvement is necessary, in order to coordinate the activities of industry based in a number of different countries. Then ETSI interpreted the EC mandate. It did not develop functional specifications to describe the radio (leaving these to IEEE and Ecma), but instead focused its efforts on the creation of test standards for UWB to ensure that a device shipped into Europe meets the regulatory requirements. By comparison, devices built for use in the USA are tested according to a methodology defined by the FCC and executed by authorized commercial test labs.

In China, government takes an even more active role in the standardization process. The Chinese government actively encourages academia and industry to develop intellectual property in advanced technologies. The government then works to create a favourable environment where Chinese industry uses this intellectual property in standards that are principally employed in China. There are a number of potential advantages to this type of strategy. If done correctly, it can accelerate industry activity and focus it into specific topics of interest.

This approach also places a great deal of responsibility on the government to identify correctly areas of potential interest in a timely manner.

China's historical experience with high licensing fees in cellular technologies and consumer electronic equipment has strongly influenced its approach to standards generally. The government viewed these royalty payments as detrimental to the growth of the Chinese economy. Royalties would take money out of circulation in China and send it to patent holders in Japan, Korea, Europe and the USA. China's approach to standardization reflects a strategy wherein royalties leaving China are discouraged and, instead, domestic intellectual property is actively encouraged. This objective has necessitated a more hands-on approach by government in standards development.

Because UWB was envisioned to be a global technology, it was necessary to develop a strategy that could accommodate these various approaches to standardization. One approach that was considered was to create a specification in a special-interest group instead of a standard in one of the recognized standards bodies. This would involve a group of companies acting in a forum such as the WiMedia Alliance to create a specification for the design of the radio. This approach was rejected. It was generally believed that standardization conveyed a greater sense of legitimacy from which it would be possible to participate in discussions with regulators and administrations around the world.

Once it was decided to pursue standardization, the next step was to select a principal path. Reaching ISO approval was viewed by participating companies as important, because ISO approval provided a more seamless method of being recognized by administrations. It also enables an industry to encourage governments to engage in discussions on its behalf as needed at an international level. While there is no specific issue that requires national-administration involvement at this time, it was viewed as a valuable asset.

7.5 Standards' role in international trade

As a final portion of this discussion on standards strategy, a little more effort will be expended to discuss the relationship between trade treaties and technology standards, as this figures into the decision to pursue ISO recognition. It is a little-known fact in the high-tech industry that there is a close association between standardization and international trade agreements. It is generally assumed (by US-based engineers at least) that a US model applies worldwide. A standard is produced and if it is well written and timely, it will be adopted and will become market dominant around the world. While this has occurred in some instances, this is a somewhat simplistic view of the situation.

Years ago, when international trade specialists were attempting to negotiate agreements to remove barriers to trade between countries, it was recognized that standards could be used as a barrier. A country who takes a more active role in standardization could easily issue laws stating that only products complying with a locally generated standard will be allowed to be sold within the country. Such a standard could be argued to be open in that all manufacturers are allowed to build products for sale within the country so long as they build to that standard.

But the process of creating standards may be skewed to favour the local manufacturer. The structure of the standards body could then be used to limit the participation of international manufacturers or to provide local manufacturers an advantage in the process. A procedural rule as straightforward as requiring that all proceedings be conducted in the local language could have the effect of limiting international participation. Local companies could then use this advantage to insert intellectual property and charge royalties or to gain a time-to-market advantage over their competitors.

Trade negotiators attempted to get rid of this opportunity to create barriers to trade by requiring that a country accept products for sale that comply with internationally recognized standards. Recent events have further narrowed the definition of 'internationally recognized standards' (in practice if not in law) to those standards approved by ISO, based upon administration-level voting..

Standards gaining ISO approval are, therefore, provided some degree of protection against trade barriers designed to create an uneven playing field in favour of the local manufacturers. Of course, this protection is not absolute. In the case of radio technologies, each country has absolute control over the use of its radio spectrum. However, as long as the standard can be employed in a manner that does not conflict with local spectrum regulations, trade treaties should apply.

In the event that local rules are discriminatory, it is possible for affected manufacturers to take their claims to their national administrations. Based upon these trade treaties, the manufacturer's national administration can then appeal to the World Trade Organization (WTO) for a change in the discriminatory practice.

7.6 Ultra-wideband in the IEEE

For those of you who have done any web-based research on UWB, you have seen names such as IEEE and 802.15.3a mentioned liberally. Although these are not relevant to the present state and direction of UWB, they are a significant part of the historical record of UWB. These groups clearly account for the overwhelming amount of press that has been generated on the topic of UWB and so a few paragraphs are added to place that information into a historical context.

In January of 2002, the IEEE approved a work authorization for the 802.15.3a committee to build an alternate physical layer for use with the 802.15.3 personal area network (PAN) standard. In non-IEEE terms, this means that the IEEE had an existing radio design that was developed by the 802.15.3 committee in 2.4 GHz. Authorization was being granted to put a different radio into the design, but to keep the MAC that was developed by that committee. The work authorization generated as part of the IEEE process did not dictate to the 802.15.3a committee which radio technology they were to use. Instead, the radio selection is decided by the committee based upon contributions and decisions.

From January until June of that year, work was done to describe the process that would be used to make decisions. Channel models were developed, which described the expected radio conditions. These models were necessary to compare the expected performance of radio proposals. In addition to the technical decisions, there were a number of procedural plans made as well. The committee developed a voting process by which to select the winning radio proposal. The process documents described every nuance down to the level of speaking order and the conditions under which a proposer would have the right to gain speaking time. To say the least, these documents were intentionally detailed to avoid confusion and debate during the selection work.

By June of 2003, there were 21 proposals from companies and academic sources that were made to be considered for adoption by the committee. This represented an incredibly large number of radios to be evaluated, ranked and prioritized. Based upon this volume, it became clear that the selection process alone might take several years to process. If this were allowed to proceed, it was highly likely that individual manufacturers would proceed directly to market without a standard and the IEEE work would be made moot.

To reduce this time to something more reasonable, a number of companies began to meet outside of the IEEE process for the purpose of merging similar proposals in an informal setting. As it turned out, several proposals used a multibanded design, wherein the spectrum was divided up into a number of different channels. It was assumed that this commonality would make the job of merging proposals easier. The merging was done under a guideline that technical superiority for the intended applications would be the sole criterion used in making the selections. Very rapidly, this group was able to reach consensus around a merged proposal.

Additional merge meetings were then held in the spring of 2003 in which other proposers were invited to participate. As with the initial meeting, proposals were evaluated based upon technical merit. It was also an understanding at this stage that if a company agreed to participate in the merge process and were not selected as the superior design, they would still support the resulting merged work. This process was extremely effective in both reducing the field as well as laying the groundwork for additional merges once the official selection process had begun.

The selection process was begun at the May meeting and by the end of the June meeting, the field was reduced to two candidates. The two proposals included a scheme named direct-sequence UWB and one called multiband orthogonal frequency-division multiplexing (MB-OFDM). As you can see, engineers lack a certain knack for catchy names.

At this point, the proceedings reached a deadlock. Almost three years of further debate and discussion did nothing to help. The two camps were not to be seduced into a more cooperative mood by tropical beaches in Hawaii and Australia nor bored into submission by solitary confinement for weeks at a time at the Dallas Fort Worth airport (where the IEEE committees met).

Sheer exhaustion and the clear futility of the situation persuaded the two groups to agree to terminate the 802.15.3a proceedings without creating a standard. In December 2005, the 802.15.3a committee voted to approve its own dissolution and the IEEE accepted that decision in January of 2006.

There have been no further efforts to standardize the direct sequence technique and there are no silicon manufacturers of which we are currently aware who are attempting to build DS designs for one of the principal markets. At this time, there are no industry efforts to build a UWB radio different from the WiMedia UWB design for applications in the consumer electronics, personal computer or mobile handset markets. There are, however, several proprietary schemes, which are being pursued by individual suppliers.

The MB-OFDM camp in the IEEE recognized that the standards process in IEEE 802.15.3a was irreparably jammed and began work to analyze the failure and to plan how to go forward with standardization without repeating the stalemate that occurred in the IEEE meetings. Leaving the IEEE was a major step in the development of UWB. The IEEE's reputation in technology standardization is extremely positive and influential. The companies involved in the decision to change standards directions was not taken lightly.

In the analysis that was developed, several factors contributing to the roadblock became obvious. First, the IEEE process in 802 committees mandates a 75% approval level for all technical votes. On the surface, this makes a great deal of sense. The drafters of this rule wanted to make sure that the engineers were in support of any changes that went into the standard. In practice, the situation was somewhat different.

Imagine an intentionally extreme example where the voting level is set at 95%. In this situation, a group that is at risk of losing a technical decision need only gain 6% support to prevent the committee from proceeding. The balance of power shifts in favour of the minority. To get around this situation, one would expect the committee to have dual (or multiple) solution standards. One would also expect to have protracted discussions until such split-standard compromises were reached.

While the 802 requirement of 75% is much lower than the 95% used in this example, the principal is the same. If the approval rate is set high enough, power shifts to the minority and one would anticipate seeing an increasing number of symptoms being demonstrated. The history of 802.11 WLAN standards of the last 15 years demonstrates this condition. They have several cases of split standards in their history as well as protracted disagreements.

The second factor that was identified as problematic was the IEEE rule of one man, one vote. Again, on the surface, this appears to be very democratic. However, it turns out to be an opportunity to manipulate the committee. If a company is willing to bring a large number of employees or is willing to hire consultants or to sponsor students to come to the meetings, it becomes possible to influence the outcome of a decision. If a company has a particularly high level of motivation, as in the case of a venture-capital-funded start-up who has a finite budget and may have already begun designing silicon, the incentive to proceed in this direction is overwhelming.

In searching for an alternate standardization path, a number of standards organizations were evaluated and discussions were held with them. In the end, it was decided that the work would be moved to Ecma International under the TG20 committee. Ecma's organizational policies forced a policy of one company, one vote, which removed the incentive to send volumes of people to committee meetings, and it used a consensus decision process, which could be backed up by majority voting if required.

The TG20 committee moved very quickly to get standardization on UWB finalized. In December 2005, TG20 completed work on

ECMA 368, which described the PHY and MAC, and ECMA 369, which described the interface between the PHY and MAC. The TG20 committee unanimously approved these standards both for publication and to be forwarded to the ISO fast-track process.

7.7 Summary

The initial effort to standardize UWB was conducted in the IEEE's 802.15.3a committee. Anybody conducting a web search of literature under this topic will find that the efforts there were both contentious and well chronicled. In the end, it was a fruitless endeavour. The work on UWB in the IEEE was terminated without conclusion. But this was not the end of the story. The ball was picked up by Ecma International, who completed standards for the physical layer, the media-access-control layer and for the interface between them. This work was later advanced to ISO for international approval.

Owing to the need for specific measurement standards in Europe, work is under way in ETSI to develop these documents. The ETSI standards don't define how the radio is to be built, but instead define only the performance levels that are required to gain regulatory approval.

The decision to seek ISO approval of the Ecma standards was taken principally because of the connection between ISO's international standing and international trade agreements. By gaining recognition under trade agreements, it is possible to enlist the aid of administrations in resolving disagreements.

References

[1] www.ecma-international.org/memento/history.htm, accessed 4 October 2007.
[2] www.ecma-international.org/publications/standards/Ecma-368.htm, accessed 4 October 2007.
[3] www.ecma-international.org/publications/standards/Ecma-369.htm, accessed 4 October 2007.

8 Special-interest groups

When a new technology is established as a standard, it gains a degree of validation. Standardization is a statement that the industry has largely consolidated its opinion around an approach for a new technology. This is important from a market-development perspective because it sets the technology onto a path of broad recognition and acceptance. Many people believe that when the standard is completed, the work of coordination in the market is done and the market will develop on its own from there.

The development of the standard is usually the first and most public of a series of activities designed to coordinate the market evolution. The work in the standards body is only meaningful when it is properly coupled with work done in special-interest groups (SIGs). These organizations insure interoperability, establish terms for access to intellectual property, manage brands, speak on behalf of the industry and generally take responsibility for the ongoing management of the market. It is not uncommon for special-interest groups to take an existing standard that has multiple modes or options, which may have been included to obtain political support for the standard, and whittle those down to the essentials. While participation in the standardization process is undoubtedly important to a company developing UWB products, participation in the SIG is at least equally so.

There is no set process through which the industry decides to structure a SIG. As a rule of thumb, a SIG is created when the industry perceives the need to coordinate the activities of manufacturers. Historically, SIG creation has occurred in an effort to accelerate slow standardization processes, in an effort by one manufacturer to gain acceptance of a proprietary design, as a mechanism to engage government with a unified voice, as a means of sharing the costs of promotion, and dozens of other reasons. In the short-range wireless space (WLAN and WPAN), the usual rationale for a SIG is to perform interoperability testing and to allow the industry to speak with one voice to government.

When a SIG comes into being, a group of companies get together and establish the legal documents which establish the group. Under normal conditions, these documents describe the purpose of the organization, its structure, membership requirements, rights and priviledges of members, sharing of intellectual property, funding mechanisms and a variety of other logistical details. In most cases, the foundational document will also establish a board of directors that makes the highest-level decisions for the organization.

Looking more specifically at the SIGs that are in the short-range wireless market in one form or another, there is the Bluetooth SIG, the WiMedia Alliance, the USB Implementer's Forum, the Zigbee Alliance and WiFi. Among these organizations, the normal structure of the SIG is to begin with the founders or promoters, who usually start the SIG and act as the board. There is then usually a second tier of companies, who come to the party later and so are not as likely to be board members. These companies usually have a fee for joining the SIG which is less than that paid by the founding promoters and have fewer rights and privileges. They normally participate in either the development or at least the review of any technical documents that the SIG produces.

The third tier of companies is the adopters. Adopter companies normally pay substantially less money for membership, and consequentially have the fewest rights and privileges in the organization. Their membership usually gains them the rights to copies of the specification, intellectual-property rights (IPR) and the rights to participate in very limited numbers of meetings of the SIG. Normally, adopters intend to employ the technology in their products, but they are usually purchasing components rather than building the capability from scratch. They wish to be knowledgeable on the technology, but not expend the resources to design it.

8.1 An overview of UWB special-interest groups

Most industries don't require more than a single SIG to represent their interests. The number of involved SIGs in a new technology would normally be one, or possibly two, in most cases. The situation in UWB is different. Ultra-wideband is a core technology that is part of a convergence between the consumer electronics, personal computing and mobile communications sectors. Each of these sectors had SIGs that were active and which independently elected to move into UWB. These SIGs then began to organize their efforts in such a way as to work cooperatively on the technology instead of competitively.

Promoting communication uses of UWB, there is the WiMedia Alliance, the Bluetooth SIG and the USB IF. These organizations will be discussed in depth in this section. In addition to this core group, there is also the 1394 Trade Association and Zigbee, UWB Forum which have the potential to become more involved in the UWB ecosystem, but which are not presently active. A more abbreviated discussion will occur on each of these organizations as well.

8.2 The WiMedia Alliance

It was mutually decided that instead of having each SIG develop its own UWB radio, the radio developed by the WiMedia Alliance would be made available to the Bluetooth SIG and the USB IF for use with their respective protocol stacks and to any other organization that might have need of the radio as well. This agreement resulted in the development of the common radio platform shown in the Figure 8.1. The two lower layers represented in Figure 8.1 are the radio. The portions above it are protocol stacks, which give the radio a 'personality' from the perspective of the user. By using a common radio, it is possible to gain economies of scale and reduced unit prices from higher-volume production. When a manufacturer of PCs or mobile phones wishes to include UWB, there is only one radio that needs to be included into the device, no matter how many protocol stacks are to be supported. It is also possible to avoid having conflicting radios interfere with each other when they work in very close proximity to each other.

Because of the common-radio-platform agreement, the WiMedia Alliance is the principal SIG developing the UWB radio design. The Bluetooth SIG and the USB Implementers Forum submit requirements

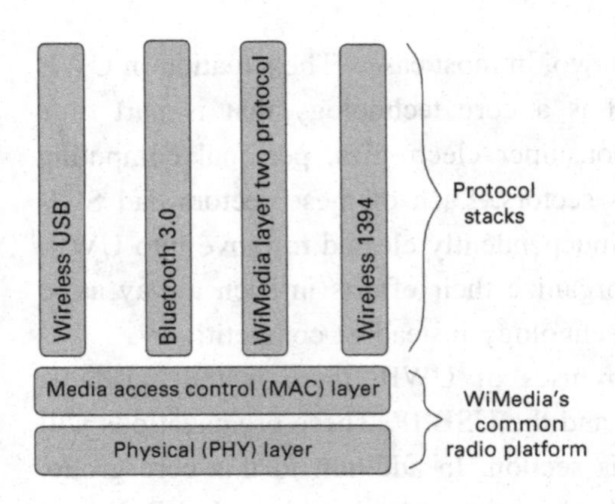

Figure 8.1 WiMedia common radio platform

to the appropriate WiMedia engineering committees. WiMedia acts as a location where companies representing various UWB applications can negotiate trade-offs.

Member companies also discuss future evolutionary directions for UWB and then work independently in a variety of standards fora to promote those views. This informal discussion process makes future standards work go more smoothly by aligning industry opinion in advance of standards voting.

At this time, WiMedia is focused exclusively on UWB radio technology. However, other PAN-related radio technologies are beginning to emerge in addition to UWB. These new technologies are still several years away, but have the potential to become relevant to future PAN requirements. It would be reasonable to expect that WiMedia will expand its charter to embrace these technologies as appropriate and to make them available to their partner organizations.

WiMedia also steps in to play a role in interoperability testing of UWB radios. When a standard is completed, it is rarely the case that manufacturers can take that standard and build it without their engineers being required to 'interpret' some portion of the standard because of a lack of clarity in the wording. What appears to be perfectly clear text when viewed in the context of standards-committee debate by somebody who has seen the evolution of thought over time, proves to be somewhat more opaque when implemented by an engineer who has not been exposed to the same discussion. When two companies interpret a standard differently, they will probably end up with incompatibilities that may prevent their radios from communicating. Because of this, it is necessary to take the new devices and test them for interoperability, to identify and clarify areas of ambiguity in the standard.

The process of interoperability and certification testing is also useful to refine the standard beyond clarifying ambiguities. When engineers design a new product, it is normal for them to be overly strict in specifying performance of the device as described by the interoperability and certification-test documents to try to ensure that it operates properly. Once these same engineers have greater exposure to the devices after they have been in the market for a while, they begin to optimize the number and type of tests they perform on a given product. Test suites become shorter. Tests that are not needed to guarantee performance are gradually eliminated. Parameters are loosened where appropriate to reduce product costs.

This type of interoperability and certification testing falls outside the responsibility of most standards bodies and so is frequently undertaken by a SIG that wishes to employ the technology. Each SIG promotes a brand such a Bluetooth, USB or WiFi, which tells customers what a product is good for and provides an assurance to the consumer of interoperability and performance. In the case of UWB, WiMedia writes the interoperability tests. This includes testing for the UWB physical layer, media-access control layer and the Internet-protocol stack.

If the final UWB product is going to bear a WiMedia logo, WiMedia will test the product using one of its approved test labs. If, however, the end product is going to bear the logo of either Bluetooth or Certified WUSB, the test suite for the layers included in the common radio platform will be given to Bluetooth and the USB IF, respectively, for execution with test labs that they have authorized. In the longer term, it is the intention of the three SIGs to employ common test labs to allow vendors to go to a single location to obtain all testing that might be required, regardless of the protocol stacks employed in the design.

WiMedia also takes on a number of additional efforts related to the expansion of the overall UWB market. This includes acting as the principal spokesperson for the press for stories related to UWB radio topics. It also includes engagement with national administrations for the purposes of creating radio regulations governing UWB. And finally, the WiMedia Alliance acts as the principal conduit by which companies are able to gain intellectual-property rights for UWB designs (discussed more later in this chapter).

8.3 The Bluetooth SIG

The Bluetooth SIG has a long track record in the PAN market since it produced a 2.4 GHz radio that was optimized for 10 metre applications. Since Bluetooth radios were originally built by companies having a strong interest in mobile phones, it should be no great surprise to find that it was optimized for voice and low-data-rate applications that are well suited to that application.

In 2006, the Bluetooth SIG reached agreement to work with the WiMedia Alliance on the UWB common radio platform. But, there are some differences in the designs that will be used. The first of these is that Bluetooth has stipulated that they will be operating their radios in the bands above 6 GHz. This is because fourth-generation mobile-phone networks are likely to be deployed in the spectrum around 3–4 GHz and the network operators would like to preserve this spectrum for their use. Employing UWB in the 3.1–5.0 GHz spectrum, as is presently allowed in the United States, would go against that objective.

The Bluetooth SIG is also taking into consideration the timing of its use of UWB. Its decision to use only radios operating above 6 GHz means that it is likely to require about 18 months longer for radio development than would be needed for radios in the 3–5 GHz range. On first blush, it would appear that taking an 18 month time-to-market hit would be a serious competitive setback. In the case of the mobile-handset market, this is not as significant as it might originally seem.

As anybody following mobile telephony is well aware, the capabilities of a handset are growing rapidly. Phones are now regularly equipped with cameras, games and MP3 players. On the near-term horizon are camcorders built into mobile phones as well as higher resolution cameras and a variety of additional functions.

At present, the data rate for 2.4 GHz Bluetooth radios is still sufficient to keep up with current communications requirements and is well matched to the existing bandwidth available on internal communication buses. However, although the amount of time required to move picture files and MP3 files on to and off of the phone is viable, there is insufficient bandwidth available to pursue transfer of large image files or video files as cellphone manufacturers would like. The introduction of a UWB version of Bluetooth is well timed to enter the market at about the time that the phone requirements exceed the capabilities of the 2.4 GHz design. If UWB were introduced on phones today, it would be overkill for most applications. In 18–24 months, when the 6 GHz version is ready, the needs of the mobile handset will grow to be a much better fit for UWB's high throughput.

It is also important to note that the Bluetooth SIG is unlikely to endorse a hard transition from the 2.4 GHz designs to a UWB radio. At the time of this writing, it is much more likely that the Bluetooth SIG will endorse a hybrid radio design where 2.4 GHz and UWB are coupled together to provide support for legacy products. It is hypothesized that the 2.4 GHz radio will be used as a control channel that can coordinate the activity of all versions of Bluetooth and that UWB will be turned on whenever a high transfer rate is required.

8.4 Universal-Serial-Bus Implementer's Forum

Just as the Bluetooth SIG concentrated in the mobile-handset markets, the USB IF has a tradition of producing products for the PC market sector. The USB IF is responsible for the promotion of the universal serial bus (USB) designs that are used to connect peripherals to PCs. Ultra-wideband will be the first time that the USB IF has attempted to engage in radio designs. But this shift from cable to radio is not unexpected.

It has been clear for a long time that consumers find the wires to connect PC peripherals and home entertainment components to be both confusing and unsightly. If one further complicates this situation by including the emerging class of mobile devices into the mix of devices which connect to the PC, the consumer is also being required to carry cables with their phone or camera to allow them to connect to PCs in addition to the ones that they already have at the home or office. This situation is quite readily managed with the introduction of radio technology. As an example, consider a camera that is equipped with a radio interface. It is possible to connect to a PC at home, the office and an Internet café with equal ease and with no new wires.

This is the situation that is driving cable replacement as a primary application for PAN radios. Cables, which exist in the PC and CE clusters, are all being targeted for replacement by manufacturers. This is the motivation that has brought the USB IF into the PAN market environment. The USB IF will attempt to address this requirement by porting and adapting their USB protocols for use in a wireless environment and will use the brand name Certified Wireless USB for this design.

The 'Certified' portion of their name has significance, which is worth noting. A number of manufacturers who have wired USB capability attempted to gain a head start on the cable replacement market by doing a direct port of their existing wired USB protocols onto either a proprietary flavour of UWB or another radio technology. They would frequently describe their products with names like 'wireless USB' or 'cable-free USB'.

The USB IF did not elect to follow this type of a path because of two principal problems that exist and are unique to the wireless domain. First, the original USB protocol is written with the assumption of extremely low error rates. In a wired environment, this is an acceptable decision. In a wireless environment, the nature of a radio connection is that it naturally encounters a much higher number of errors even on a very good link. To make the USB protocol work well in a wireless environment, it needed to be made robust against these errors.

The second major problem that is unique to a wireless environment is the issue of security. In a wired environment, permission to connect to the PC is implied by the person taking the cable and plugging it into the device. The cable length is relatively short and it is assumed that anybody attempting to eavesdrop or penetrate a computer would be readily visible to the authorized users. In a wireless environment,

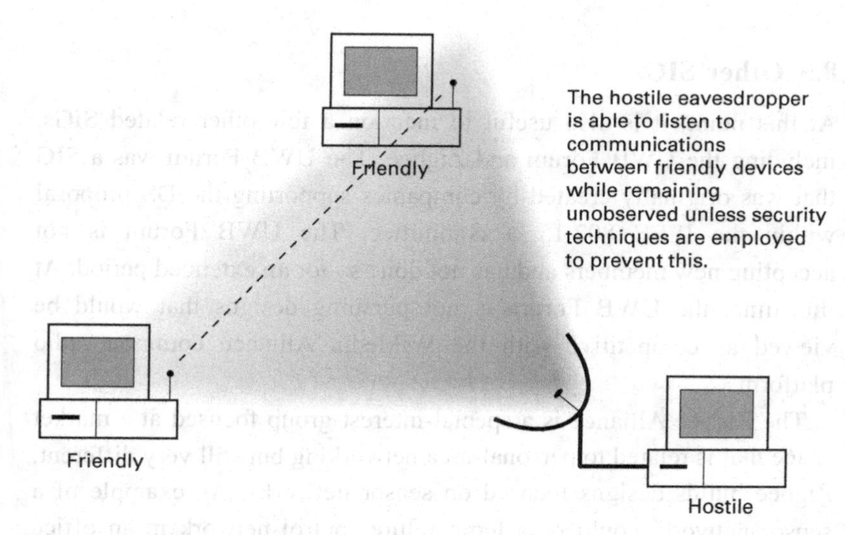

Figure 8.2 Wireless security risks

eavesdropping and unauthorized access are not nearly as detectible by an authorized user. If no changes were made to the USB protocol as it was moved into a wireless environment, eavesdropping could be achieved by using little more than a directional antenna. Figure 8.2 tries to demonstrate how this can be done. The hostile PC, with the use of a highly directional antenna, has the ability to stand off from the legitimate conversation and monitor the traffic. The directionality of the antenna allows it greater separation than a normal consumer would experience. This makes it less likely that an authorized user would notice the eavesdropper's presence.

If the individual PC can be compromised in this way, it is also possible to use the compromised PC to act as a gateway into the larger network to which the PC is attached. Because of this, it was simply unacceptable to the USB IF and its customers to leave such gaping deficiencies in the design. This caused the USB IF to restructure the USB protocol in a manner that fixes the problems but still keeps the ease of use which USB customers find attractive. And to differentiate their design from those vendors who are simply porting wired protocols, the USB IF will call their design Certified Wireless USB.

8.5 Other SIGs

At this point, it is also useful to mention a few other related SIGs, including the UWB Forum and Zigbee. The UWB Forum was a SIG that was originally created by companies supporting the DS proposal within the IEEE 802.15.3a committee. The UWB Forum is not accepting new members and has not done so for an extended period. At this time, the UWB Forum is not pursuing designs that would be viewed as competitive with the WiMedia Alliance common radio platform.

The Zigbee Alliance is a special-interest group focused at a market space that is related to personal-area networking but still very different. Zigbee builds designs focused on sensor networks. An example of a sensor network would be a temperature-control network in an office building. Thermostats are distributed throughout the building and are used to control air ducts for heating and cooling. The cost of running cable to all of these thermostats is prohibitive. So Zigbee is designing a radio network to serve this purpose. In a Zigbee network, the sensors (e.g., thermostats) are most often battery powered. The network topology needs to be self-organizing and capable of recovery from failure of individual nodes. The amount of data transferred is relatively small and the frequency of data transfer is relatively low.

While it would technically be possible for a sensor network to employ UWB, this is not necessarily the best radio for the task. Some work has been done in the IEEE 802.15.4 committee to define UWB for these uses, but the majority of activity is focused on 2.4 GHz radios that have greater range and penetration capability.

8.6 Special-interest-group (SIG) operations relating to UWB

When the last standard for any radio technology is approved after years of debate and the pen is finally put down, a great deal of work remains to be completed before a consumer will be able to walk into a store and purchase a printer with the confidence that it will work with the existing PC or camera at home. This is where the special-interest group emerges to fill the gap. These organizations are responsible for the interpretation of the relevant standards to make sure that the resulting devices can work together. They are responsible for coordinating access to intellectual property rights to enable manufacturers to build products. They are also the principal bodies which direct the development of the industry.

In most sectors, there are one or two active SIGs. Normally, when there is more than one working in a common space, the SIGs are in competition with each other for the right to direct and represent the industry. In the PAN market, there are substantially more SIGs than normal. Much of their activity is cooperative, but some competition is unavoidable. The Bluetooth SIG, the USB IF and the WiMedia Alliance have all agreed to work cooperatively to develop and deploy the same radio so that consumers may avoid the inconvenience of interference and benefit from the economies of scale. Manufacturers can benefit by being able to deploy one radio to serve several applications.

But the PAN space is a convergence between the PC, CE and mobile-handset markets and, because of this, the customers and applications addressed by the SIG will become increasingly common. The Bluetooth radios are beginning to appear on PCs to enable wireless mice and keyboards in a market that has traditionally been the domain of the USB IF. Likewise, USB connectors are beginning to sprout on some mobile phones to enable them to connect more readily to the PC for synchronization operations. In this regard, there will be some give and take as the industry seeks a new equilibrium. Because the interaction of these organizations is expected to remain in flux for a number of years, product development teams must establish an understanding of each of these SIGs and the way that they operate. Specifically, it is valuable to understand how these organizations treat IPR, interoperability testing and logo rights.

8.6.1 Intellectual-property rights

Before diving into the detail of how the major SIGs handle IPR, a brief primer on the topic is warranted. This abbreviated discussion is not intended to be a substitute for legal advice. Neither of the authors is a lawyer nor do we aspire to spend more time than absolutely necessary in conversation with lawyers who are operating in a professional capacity. But having done so across our chequered past, we will attempt to relay a few pointers.

The single most important thing to know is that no matter how hard you try, it simply is not possible to create a bullet-proof shield against intellectual-property claims as long as you choose to build products. The best that you can hope to accomplish is to minimize the number of companies who are in a position to make claims against you or to maximize your ability to negotiate when a claim is presented. Special-interest groups are a major tool in trying to establish protection for your organization.

A SIG will normally have an intellectual-property license agreement built into its membership documents. These documents will vary substantially in the detail, but at the highest levels there are very few IPR permutations. The SIGs operations in personal-area networking use either reasonable and non-discriminatory (RaND) terms or reasonable and non-discriminatory with zero royalty (RaNDZ) terms. In the case of RaND terms, the members of the SIG are agreeing that you will be allowed to use their intellectual property. The terms under which use will be permitted are not stated in the agreement. Royalties and scope-of-use limits are permissible under this system. The licensing companies are simply assured that the patent holders will not take the step of telling you to cease production; however, this also does not mean that royalties will be assessed. In the overwhelming number of cases in the PAN markets, they actually operate as a zero-royalty system even if that is not legally required.

By contrast, the RaNDZ terms do provide a legally binding system where no royalties will be charged. You will not be charged a royalty by other SIG members under the RaNDZ agreement. This statement substantially simplifies the agreement terms and overlooks a lot of detail. There are usually additional conditions applied, such as defensive-suspension and essential-patents clauses, which modify your rights and obligations. Although these clauses are material, they are second-order effects. These are beyond the scope of this document to describe. If you are concerned about this level of detail, please consult an attorney.

At a simplistic level, RaNDZ terms appear to be highly preferred. It makes the world a simple and safe place to know that you won't be charged royalties. It's very straightforward and easily understood. Unfortunately, life is fully of little complexities. From the perspective of a large company with a broad offering of products, RaNDZ may not be the most desirable approach. Say, for instance, that you have a core business that is not related to the SIG work and a secondary business that is covered by the SIG agreement.

A company competing with yours violates your patents in the core-business area and your company wishes to respond. The normal process for negotiating such a violation is for each of the two companies to create a stack of the IPR that the other company is believed to have violated. Your company will attempt to claim enormous damages, which must be compensated. The opposition will attempt to claim that there are offsetting violations, which neutralize your company's claims. It may even prove to be the case that your company has the greater liability.

For medium and large companies with broader product lines, it is not unusual to find that they prefer RaND terms. Imagine a situation where a company (Defender Corp.) has two product areas. One is mission-critical to the business and the second one is less critical to the Defender's success. The SIG agreement covers intellectual property related to the secondary business. In a defensive situation brought on by infringement claims from Intruder Corp., Defender wishes to bring as many counter claims as possible against Intruder to provide a better negotiating stance. If Intruder has products that are covered by the SIG license agreement, Defender would use the flexibility of a RAND license to request royalties for any patents which Intruder has used. The two companies would then negotiate out any actual licensing agreement by weighing the relative strengths of both sets of claims.

In an offensive action, Intruder begins to build products in Defender's principal business area. Intruder's core business is in the business sector where the SIG is operational. While Defender has significant IPR coverage in its core business, it wishes to increase the degree of pain inflicted on Intruder to raise the stakes and to warn off others who might emulate Intruder's expansion. In this case, Defender would again bring the patents covered by the RAND license to the table and request royalties there, as well as in Defender's core business.

The Defender–Intruder scenario is most likely to occur between smaller companies. When larger companies are involved, their patent resources are usually deep and their legal resources extensive. It is not usually profitable for a small company to take on this type of an adversary unless its position is unassailable. When large companies engage each other, the patent coverage that they have creates an extensive web. If another large company engages them, it is highly probable that the number of products affected in the claims and counter claims will grow to be quite large. The costs of such actions are very high and the potential for debilitating judgements is very high as well. This risk establishes a mutually assured destruction (MAD) approach to intellectual-property rights. Anybody initiating action is almost assured of being badly injured. While this sounds very hazardous, in practice, it results in a very stable system. It is generally unprofitable to initiate patent actions.

To make matters just a little more complicated, corporate strategies related to IPR change by industry and over time. Several years ago, RaNDZ was reasonably accepted in the communications markets. In the last few years, RaND has been preferred. In the computer industry, royalties have been largely discouraged in the belief that royalties tend to discourage market growth by raising customer prices. In the consumer electronics markets, royalties have been an accepted fact of life that is generally not believed to affect consumption.

In the PAN space, there is a mix of IPR offerings. Bluetooth is a RaNDZ SIG. WiMedia and Wireless USB are both RaND SIGs. In all of these cases, royalties are either very low or non-existent. The principal fact, which appears to determine whether royalties are charged by the major players, is the expectation established at the creation of the SIG. If the major players wish to pursue royalties, then structures will be created to enable and encourage royalties to be sought. Conversely, if the major players do not wish to pursue royalties, they put into place structures and expectations to discourage the practice.

The element that cannot be controlled is the unaligned IPR holder. Companies outside the reach of the SIG are not bound by the terms of the agreement and are free to charge royalties if they choose. The MAD terms used by the major companies in a RaND environment work reasonably well to suppress this for unaligned product companies. If these companies have products, they are most likely infringing upon IPR somewhere. Starting a fight with large companies on IPR is an expensive and protracted proposition that most product companies cannot afford.

The most unreachable IPR adversary is the troll. These are companies whose purpose in life is to collect IPR and to seek out companies who are violating it for the purpose of charging royalties. Because they do not have products of their own, it is difficult or impossible to negotiate with them. The most effective way to combat charges of this nature is to try to get the patent invalidated. Once again, this is a difficult and expensive task. For small companies, these trolls act like parasites. They are annoying, but not usually lethal. Fighting them is more expensive than cooperating. Because of this, they are usually tolerated.

Larger companies have lawyers on staff, which makes the cost of defense against trolls more manageable. Larger companies also have deeper pockets, which tend to attract trolls more vigorously. Because of this combination, larger companies are more likely to resist royalty payments to trolls. Obviously, each situation will warrant a unique response.

There are two rules of thumb that can be derived from all of this. First, either RaND or RaNDZ will work very similarly in practice. What are relevant are the intentions of the major participants in the pursuit of royalties. Second, you probably cannot defend against trolls and unaligned players unless you are a major corporation. Engineers and marketing people should worry about IPR attacks just enough to keep questionable IPR out of specifications and standards. Beyond that, it is normally a waste of time that could be better spent developing new technology.

In addition to knowing what the IPR conditions are for an organization, it is also necessary to know when a company gives IPR rights to other members and when it receives IPR commitments in return. These events are not necessarily simultaneous. In the case of both Bluetooth and Wireless USB, a company commits their intellectual property in any existing specifications at the time that they sign the members' agreement. The other members of the SIG likewise give intellectual property rights to a new member at the point that the new member joins the organization.

By contrast to this, WiMedia has taken a different approach. Like Bluetooth and WUSB, new members commit their intellectual property to the other WiMedia members for any existing specification at the time that they join the SIG or when a new specification is approved. But members only receive IPR rights at the point that they build and successfully certify a compliant product.

To explain why WiMedia took this path, an example might be useful. Say, for instance, that WiMedia had a policy that provides members IPR rights to any specification that they generate that are awarded at the time that membership documents are signed. Since WiMedia produces work in both the MAC and PHY layers, this would mean that the new member has rights to each spec used separately or together.

It would be possible for the new member to use the rights to produce a physical layer that is compliant with the existing standards, but to employ a completely unique MAC layer that is incompatible. The effect on the market would be to have the new member produce a radio that uses the same spectrum but may not work cooperatively with devices that are fully compliant to WiMedia specifications. The new devices could potentially act as an interference source for WiMedia-compliant systems while using intellectual-property rights from the very companies being harmed to enable the action. Instead, WiMedia took a more complex intellectual-property path, which requires that a finished product be composed of compliant elements and behaves in a cooperative manner with other devices before IPR rights are granted.

8.6.2 Interoperability and certification testing

The second major function that special-interest groups provide for the industry is interoperability and certification testing. When any marginally complex standard or a specification is completed, it is extremely unlikely that the engineers will have considered every possible contingency. It is equally unlikely that all of the developers will have read the text and extracted a common interpretation. Inevitably, there will be discrepancies. Engineers will implement a common specification in a manner that prevents the devices from working correctly because the text is vague or silent on specific points. Maybe the specification failed to consider certain operational states. Regardless of the reason, the result is that devices produced by different manufacturers will frequently not interoperate. Somebody has to test to ensure interoperability.

A SIG will perform this task on behalf of the industry and will normally communicate its approval to the customer in the form of a logo. Logos are commonly used to denote interoperability. Customers have been conditioned to understand that two devices who have similar connectors are labelled with a common logo and should expect that the two devices will communicate. Because of this, SIGs normally establish certain tests that manufacturers must pass before they are allowed to use the SIG's logo. In this way, the SIG manages the customer's experience. Companies producing poor-quality devices will not be allowed to create a negative impression for the SIG's other member companies.

In developing these tests, the SIG will normally go through a series of steps. First, they will write up a manually operated test procedure. The object here is to do something that can be executed quickly to get products into the market as rapidly as possible. It is also common practice for SIGs to begin certification only after several vendors have completed products. This is to ensure that interoperability problems that may be latent in the specification are uncovered and repaired. In the case of Bluetooth, this requirement is written into the organization's process documents. Wireless USB and WiMedia have less specific requirements, but behave in much the same way in practice.

Once the manual certification tests have been developed, the SIG will normally begin the process of automating the testing. This significantly improves the reliability of the test results by removing human error. If done correctly, it also allows testing which is performed at different labs to obtain consistent results. This approach means that a SIG can scale the number of devices that it can test in a given time and also allows it to perform the testing at geographically diverse labs, which are more convenient to manufacturers. Each of the major PAN SIGs is pursuing a similar path with regard to its testing strategy.

In UWB, there are some additional nuances to the game. Because of the way that the industry has been structured with WiMedia designing the underlying radio, a certification-test suite is being written by WiMedia to run against the physical and media-access control layers. This test suite will then be given to Bluetooth and Wireless USB to bundle with their upper layer protocol tests. WiMedia will also bundle their PHY and MAC test suite with an upper layer protocol stack. Each of the three SIG organizations will then execute both the WiMedia radio tests as well as their own upper-layer protocol tests to determine final compliance.

8.6.3 Membership rights in SIGs

WiMedia, Bluetooth and WUSB all have a requirement that companies become members before they are allowed to use the SIG's logo. In all three SIGs, it is possible to establish an adopter-level membership at a minimal cost to gain this benefit. However, if a company wishes to participate in the development of future specifications as a means of staying informed about or driving the evolution of the technology, a higher membership level is required. In this regard, each of the SIGs varies in its structure and the rights that are available.

In WiMedia, there are three levels of membership possible. The highest level of membership is the promoter. These companies participate as members of the SIG's board of directors and are responsible for the overall direction of the organization. Companies wishing to participate at the promoter level may request to do so, but a supermajority of existing board members must support the decision before a new member is added. Companies participating at the promoter level also have all of the rights of members at lower levels that are described below.

The second level of WiMedia membership is the contributor. Companies at this level are allowed to participate in technical committee meetings and at face-to-face events where specifications are drafted and future directions are discussed. Contributor companies are also allowed to join email reflectors so that they might monitor communications in various technical subcommittees. Even if these contributor companies are unable to participate actively in the technical committee work, they are able to track the development via the reflectors. For companies who wish to be involved in the specification-development work, this is the level of membership that one would wish to establish.

The third level of WiMedia membership is the adopter. The title says it all. If a company is interested in employing WiMedia designs, passing certification testing and displaying the WiMedia logo, but is not interested in sending engineers to help write the specifications, this is the appropriate level. Adopters also receive communications from the SIG that are directed to all members.

There is also a hybrid membership class known as the extended board. Companies who are in the contributor class may request admission to the extended board. The board must approve these requests by a supermajority. When approved, these companies are then able to sit in on most board meetings and are permitted to participate in debate. They are not permitted to participate in board decisions.

In general, WiMedia will employ technical committees composed of its contributor and promoter members to build specifications. But this is not the only method by which this can be accomplished. In the case of 60 GHz, the technical work was handed over to Ecma International to develop in its standardization processes. When those standards are completed, WiMedia will have the option of recognizing the Ecma standard. If this were to happen, WiMedia would use its technical committees to develop interoperability-test suites and its marketing committees to promote the new radio. This approach of referencing other standards work is a means of acquiring necessary technology without encouraging redundant development.

By contrast to WiMedia, the organization of the USB IF and the manner in which it generates specifications is substantially different. The USB-related standards normally involve two different legal bodies: the USB IF and the promoter group. If a new specification is needed, Intel invites a group of companies to participate in the development. In most cases, a draft of the specification is circulated by Intel to begin the process, but is then refined and modified in cooperation with the other promoters. When the promoters have completed a new specification, it is then offered to the USB IF. The USB IF board then votes on the acceptance of the new specification. If approved, the USB IF then develops certification and interoperability testing and begins to promote the specification to the developer community.

But this is not the only mechanism that the USB IF has to generate specifications. It is also possible for the contributor members of the SIG to generate a specification through the work of the technical committees. When complete, the board of directors approves the specification for use.

Like WiMedia, the USB IF also references outside standards and specifications when needed, to avoid duplicate development. The USB IF's relationship to WiMedia to acquire WiMedia radio technology is an example of this in practice.

The USB IF's membership levels are also very similar to those used in WiMedia. Promoter members act as a board of directors. Contributor members participate in the technical development. Adopter members use the resulting specifications, but do not participate in the development.

The noteworthy differences between Wimedia and the USB IF are in the extended board and the promoter group. The USB IF does not have an extended-board structure. Board of director meetings are the exclusive domain of the USB IF promoters. The second major difference is in the use of the term 'promoters'. In the USB world, there are actually two promoter groups. The USB IF promoters are functionally equivalent to the WiMedia promoters. However, the promoters who develop USB specifications and whose membership is based upon Intel's invitation are legally independent of the USB IF. There is no equivalent for this structure in WiMedia. This group is brought together exclusively for the purpose of developing the specification and generally ceases operations when the specification is accepted by the USB IF board.

Likewise, the structure of Bluetooth is remarkably similar to the other two SIGs. It comprises three membership levels: promoters, associates and adopters. Like WiMedia and the USB IF, promoters make up the board. When a board seat becomes available, associate members may be invited to apply for the position. Approval of a new candidate must be by unanimous vote of existing board members.

Associate membership is required to participate in Bluetooth's technical committees and to attend face-to-face meetings of the organization. Adopter-level membership enables a company to participate in certification events, to use the logo if certification is successful and to gain intellectual property rights. These rights are all essentially equivalent between the three SIGs.

Also like WiMedia and the USB IF, the Bluetooth SIG has occasionally referenced outside specifications as a method by which to avoid duplicative development. Bluetooth's relationship to WiMedia is one example of this. The use of near-field communications (NFC) that was developed in Ecma and may be used for association is another example.

For a definitive description of the structure, rights and privileges that are available in each SIG, the SIG's membership agreements and bylaws are recommended reading. Web addresses for the documents needed to understand each of the SIG's membership structure are provided in the appendix.

8.7 Summary

The PAN environment is at the centre of a convergence between the PC, CE and mobile handset markets. This means that the technology employed by these groups is becoming common. It also means that the devices produced within each of these sectors will ultimately need to be able to communicate with devices from other sectors. To facilitate this evolution of the market, the WiMedia Alliance, the Bluetooth SIG and the USB Implementers Forum agreed to work together to develop a common radio, which would be shared between them. Each group would add its unique protocol layer to the radio to preserve the personality with which consumers had become familiar.

Each group would also promote the radio to consumers under their own brand names. While the WiMedia brands have yet to be launched, Bluetooth and USB are well known by customers and billions of units of each had already been shipped. By extending these brands, it is possible to gain faster market penetration and consumer acceptance.

Understanding which special-interest groups are active in the market is only part of the story. It is also worth knowing what services these groups provide for the industry. Usually, SIGs that work in the PAN and LAN markets focus their efforts around interoperability testing and intellectual-property rights. When a standard is approved, it is not guaranteed that everybody will interpret all of its dictates identically. When they do not, it is possible that two devices will not work together. The SIG takes a standard and establishes testing to make sure that any differences of interpretation are resolved before a customer experiences a problem.

In intellectual-property rights, the SIG acts to set the terms for access to essential patent rights. In the PAN markets, rights are normally granted under reasonable and non-discriminatory terms or zero-royalty terms. While there are material differences between these two approaches, in practice, the results are frequently similar.

9 Ultra-wideband business issues

The decision to implement a UWB radio into a product is not simply a technical selection. There are a substantial number of business issues surrounding the selection of which one should also be aware. Intellectual-property obligations, price expectations and market development directions are all matters that intimately affect the potential for a successful outcome. This chapter is intended to highlight some of those issues and provide a distilled assessment.

9.1 Expected changes to the technology over time

As is the case with all technologies, UWB will evolve over time. What it is today is not what it will be in the three-year life expectancy of most computer products. In making product decisions, it is never enough simply to look at the way things are now. It is also necessary to look at trends that will occur over the expected life of a product. As an example, if one were to compare a UWB radio with an 802.11n radio today, the result will be far different from a comparison that will occur in the next three years. This section discusses some of the trends now visible, which will change the functionality of UWB.

9.1.1 Planned development in UWB

In addition to the trends that are happening as a result of general economic and market conditions, there will be focused efforts within UWB standards organizations and SIGs to evolve UWB as well. There are two primary directions along which UWB will develop. To support large-file transfer and video applications, development will push the transfer rates of UWB to the highest levels possible. There will also be a parallel effort to reduce the power requirements of UWB in order to support the use of UWB in battery-powered platforms. Both of these paths will be pursued simultaneously.

The evolution of UWB toward higher throughput is probably the more visible of the two development paths because it will enable new applications. Multi-gigabyte movie files can take an unacceptably long period to transfer even with first generation UWB's very high data rates. Increasing these data rates translates into reduced customer wait times. This, in turn, enables new application scenarios, such as a movie kiosk. The kiosk model suggests that mobile platforms such as phones or personal digital assistants (PDA) will be able to pick up data from a kiosk (or other network connection) and transfer the data for use in a disconnected environment, such as a car, plane or, train.

The first generation of UWB radio designs does not use all of the tricks available to radio designers to gain additional throughput. There is no effort to utilize higher-order modulation, channel bonding or smart antennae to increase the throughput of UWB. All of these techniques have side effects, such as increased cost, reduced range or reduced device density. While they may not necessarily be universally applicable to all of UWB's intended applications, it is quite probable that they could be legitimate options. For instance, in short-range file-transfer-type applications, sacrificing some of the range to gain additional throughput may be a legitimate trade-off.

All of these techniques have been implemented in existing WLAN designs, and so are well understood. They were not employed in the first generation of UWB as it would have added unnecessary complexity to an already challenging design. In later generations, it should be expected that several of these techniques will be included. If this occurs, it is probable that UWB will be able to increase throughput to somewhere near 1 Gb/s almost immediately, using higher-order modulation. A 1 Gb/s throughput will be included by several manufacturers in an effort to differentiate their products. In the next 4–5 years, it is quite probable that UWB will be able to support 2–5 Gb/s in the second or third version of the standard.

Just as UWB can be expected to employ high-performance techniques originally developed for the WLAN markets, it is also possible to employ power-saving techniques that were developed for the mobile-phone industry to make UWB much more power efficient. Future generations of UWB will implement techniques to allow the radio to save power by shutting down in whole or in part when inactive.

Reducing the power consumption of UWB will enable battery-powered platforms either to extend their battery life or to add additional functionality into the platform. Given the historical trend to integrate additional features such as games, cameras and navigation into these devices, it is necessary to make each component of the system as power efficient as possible to allow that integration to occur.

9.1.2 Multiple-radio integration

The devices which employ UWB radios will also experience a longer-term trend related to radio integration. Personal computers, mobile phones and consumer electronics equipment will gradually begin to incorporate multiple radios into their designs. For instance, PCs can be expected to incorporate LAN radios such as WiFi, location radios such as GPS, PAN radios such as Bluetooth 2.4 GHz and UWB as well as WiMax and other WAN radios. As the number of radios increases, there is also an increasing need to reduce the cost of these devices as well as the physical area required to include them into a device. Cost, power and space restrictions will force integration between dissimilar radios to be pursued over time. This task is more challenging than the integration of a single radio and so can be expected to take a greater amount of time to reach market. The introduction of multiple radio chipsets such as Bluetooth/802.11 combined chips is an example of this trend line that can be expected to continue.

9.1.3 Converging the WAN, LAN and PAN networks

If one imagines for a moment that a handheld platform exists that potentially contains WAN, WLAN and PAN radios, as suggested by the above trends, it makes sense that the consumer would prefer to see these operating in a fairly seamless way. The consumer wants to be able to take advantage of the functionality, but really doesn't want to have to understand a great deal about the limitations or operational details of the radios to make them perform. If this logic holds, it is likely that we will see a gradual blending of WAN, LAN and PAN networks. A consumer will want to start a voice call on an inexpensive LAN or PAN connection when it is available and migrate to a WAN network on getting up and leaving the building.

Getting this to occur will not be quick or easy. It will be a gradual process, where a lot of incremental improvements are made. Initially, this will appear as LAN and PAN hot spots emerging within a WAN carrier's suite of services. The first examples of this in the WAN and LAN spaces are occurring today, with telecommunications companies such as Vodaphone and Verizon deploying WiFi hot spots.

Gradually, it is likely that these independent networks will blend not only at a technical level, but at a business level. As an example, it is possible that a layered-data service could occur wherein headlines and story summaries are made available on a WAN service, but detailed stories and extensive photographs are not provided until the subscriber comes in range of a LAN or a PAN hot spot, where greater bandwidth can be made available. Obviously, a variety of technical hand-off, billing and other business issues will have to be resolved before fully merged networks can occur.

In the LAN and PAN portion of the market, there will also be some visible signs of blended networks. The evolution of home networking is an example. In home networking, the principal objective is to create a system that is capable of connecting smart devices around the home such as PCs, set-top boxes (STBs), personal video recorders (PVRs) and mobile phones without adding new wires. Radio-frequency networks, such as LANs and PANs, are ideal candidates for this application. These networks would principally be used to transfer relatively low data rates in the form of Internet exchanges or photographs, as well as much higher-rate traffic, such as high-definition streaming television.

However, the bandwidth requirements for multiple simultaneous streaming video channels will be extremely taxing for any single wireless network, such as a WLAN. It is quite possible to architect the home network in layers to resolve this problem. A backbone layer transports traffic when it is necessary to cross the home. If a video stream is being sent from the living-room consumer-electronics cluster to a television set in the bedroom, this would be an example of traffic that needs to cross the backbone. But not all traffic needs to cross a common backbone. Traffic such as data from a PC to a printer or between devices within the CE cluster may be segmented into separate hot-spot networks.

A hot-spot layer provides dedicated capacity for isolated portions of the network that have unusually high traffic densities. Look at the CE cluster to see how this might be employed. A CE cluster may include a television, a set-top box, a DVD player, a personal video recorder (PVR) and a game console. All of these devices may physically be within 1–2 metres of each other. Several video and audio channels may be running simultaneously. The overwhelming volume of traffic in the cluster goes between these devices. A hot-spot network like UWB could be used in that very small area to offload this localized traffic from the backbone.

Obviously, the addition of multiple layers to a network has a cost in terms of complexity and expense. If the backbone has sufficient bandwidth to handle all traffic in the home, one would clearly choose to use the backbone exclusively. But if the backbone bandwidth is insufficient to handle all of the traffic, this two-layer structure provides more network capacity. Obviously, this isn't intended to be an exhaustive discussion of home-networking architectural trade-offs, but is instead a demonstration of how a blended network might emerge in a home context.

9.2 Business and market trends

The trend lines that were mentioned in the prior section were largely related to technical evolution. There is a series of equivalent business and market related trends, which are worth understanding as well. This is not to say that these topics are divorced from technical issues. It is simply a business view of the data, which can be used to generate a different perspective.

9.2.1 Price erosion

The long-term evolutionary changes described above, such as network convergence or home-network evolution, are all predicated on the existence of an appropriately priced UWB network. If UWB were to be able to operate at 2 Gb/s and had power consumption at a level of 0.01% of the current level, but the chips cost $1000 each, they would not be useful for any of the anticipated applications. The cost would be simply prohibitive. To play any part in the evolutionary trends, UWB must experience price erosion that makes it market viable.

When UWB was first developed in the 1960s and 1970s, a transceiver cost thousands of dollars and was only reasonable for use in military applications. The first step in the price reduction path was to move from discrete components to semiconductor devices. This became possible in the late 1990s. For UWB to reach its maximum penetration into consumer devices, such as the PC, mobile handset or portable consumer electronics, the price must fall further yet. It is commonly believed that there is a 'magic' $5 price point at which the market will activate aggressively.

Historically, the majority of cost reduction occurs in the first three years after initial products hit the market. The Figure 9.1 is from Bluetooth's market introduction, which should be representative of the price decay curve that UWB will experience. This is where the technical improvements are being developed. Price reductions beyond three years will occur gradually when they are a result of general manufacturing

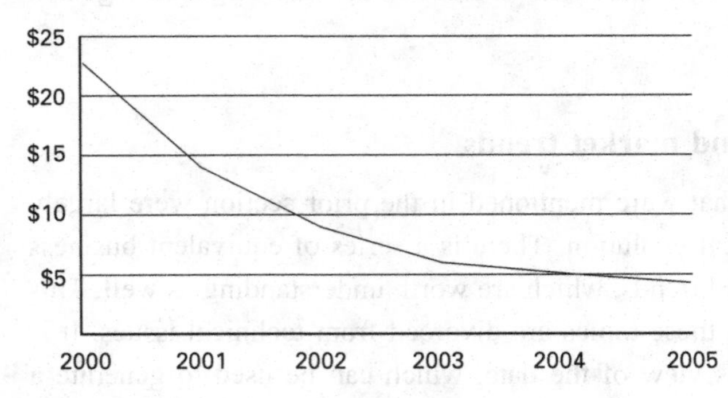

Figure 9.1 Bluetooth price decay

efficiency and reduced margins or they will occur in a more stepwise fashion if they are a result of silicon-process improvements.

9.2.2 Consolidation

Another normal trend in the early development of the market is consolidation in the number of manufacturers building chips. At the launch of UWB, it is probable that there will be a large number of chip manufacturers participating. Many of these are venture-capital-funded start-ups. This number will decline sharply during the early market. Some of the small players will be acquired by large companies who are entering the market too late to develop their own solutions. Some small companies will find that they are not competitive and will fail. Some large companies who are attempting to build chips will find that it is more cost-effective to buy components instead. All of these factors will take a toll on the manufacturing ranks during the early stages of the market. This is a normal process and should not be concerning.

If the market volume for UWB chips develops rapidly, it is more likely that the total number of manufacturers will remain relatively high. If the market evolves in a manner where the ramp begins relatively slowly, then the degree of consolidation is likely to be higher. The present expectation is that the market will be very high volume with a relatively quick market ramp. There are several factors that come together to support this belief. First, is the overall size of the market. If one looks simply at the volumes of components consumed in Bluetooth and wired USB products, there are approximately two billion units installed in each. It would not be appropriate to assume that complete cannibalization will occur, but even without overwhelming cannibalization, there is a substantial unit volume.

Second, there is brand familiarity. When a new brand enters the market, it takes a while for customers to understand what the brand stands for in terms of new capability and quality. But, when a second-generation product is introduced into the brand family, the customers are already familiar with the value proposition and are more willing to purchase. In the case of UWB, the technology will be marketed as part of the USB and Bluetooth brands in addition to a new WiMedia brand. Both of these have a long history with customers, which should help accelerate the rate of adoption.

The intellectual property implemented by WiMedia will probably be employing a new brand name and brand promise. The volume of chips sold under the WiMedia brand will probably ramp more slowly than either those under the Bluetooth or WUSB brands because of the unfamiliar brand name. However, as the radios for all three brands employ common chips and differ only by the software loaded onto them, the chip volume is additive and should be very large.

The third factor related to the growth rate and overall size is Metcalf's Law. Metcalf's Law states that for every *n* devices attached to a network, the value of the network (the total number of connections that are possible in the network) is *n* squared. In this case, by allowing multiple devices to communicate together, the value of the individual devices increases. A phone is not simply a phone. The wireless connection that a PAN provides makes it more convenient to use the phone as an MP3 player when songs are probably stored on PCs. The high-speed wireless connection makes it convenient to use the phone as a high-resolution digital camera. The value of the mobile phone as a camera is increased when it becomes easier to offload the resulting pictures for storage or modification. As the number of possible connections increases, more applications emerge, there is more value to the consumer in connecting and this, in turn, fuels additional sales of radio components. Taken collectively, these factors strongly suggest the potential for a large and fast-growing market supplied by a variety of manufacturers.

9.2.3 Rollout expectations

A lot of press, some analysts and more than a few manufacturers have asked for a better understanding of how UWB is expected to roll out. Would it be reasonable to have it explode on the scene (wishful thinking by VC-funded start-ups) or will it be a long, drawn-out process, like WiFi was. To understand how UWB will roll out, there are a handful of variables that will have a bearing on that question.

First, the value proposition to consumers has to be understood from the perspective of each product that may consider implementing it. The cost of UWB has to be weighed against the perceived value of the feature. If the device under consideration is not mobile and does not have an inclination to communicate with mobile devices, then the mobility that UWB provides will be of little to no value. Likewise, if the cost of the device under consideration is very low and the price of the UWB radio is comparatively high, it is likely that UWB will be implemented in a small volume while the chip price is high and in greater volumes as the price declines.

Taking these two factors into consideration, it seems reasonable to anticipate that some of the first products likely to employ UWB will be relatively expensive devices that derive value (i.e., mobility or cable replacement) or support mobile devices. Manufacturers of PCs are moving rapidly to deploy UWB both in support of communications with mobile devices and also for wireless docking applications, which eliminate cable clutter.

Some peripheral devices, which connect to the PC, are also moving relatively quickly as they see value for removing cable clutter. External hard drives, printers and scanners all seem to be in this class. Smaller devices, such as cameras and music players, are expected to move in more of a second wave. Many of these manufacturers wish to see a volume of devices in place, to which they can talk (think PC).

The stationary CE devices are unlikely to employ UWB to transmit streaming video for a variety of reasons, but are likely to use it to support mobile devices. This is likely to be a third wave of participants. While the mobile devices are waiting for the PCs to be relatively well established, the CE devices are likely to wait for the mobile devices to have a presence as well.

The second major factor in judging rollout is customer familiarity. For technologies that are unfamiliar to consumers and integrators, the rollout is more protracted. A brand name and value proposition need to be established and recognized before word of mouth can develop. Normally, this can take several years to establish. In the case of UWB, this ramp is likely to be a good deal sharper. By establishing relationships with Bluetooth and the Universal Serial Bus Implementer's Forum, UWB will piggyback on the well known Bluetooth and USB brand names. It is expected that this will accelerate adoption rates significantly through this means.

The third factor to consider in judging the rollout rate of UWB is the competitive presence of WiFi. WiFi (WLAN) radios exist in the market in high volume and with a relatively low price point owing to the technology's mature state. If integrators elect to support a single radio for cost reasons this will cannibalize some portion of UWB's potential market share. As this is a competitive situation, several factors may weigh on the integrator's decision.

WiFi is mature today. If the integrators' applications are not well served by today's performance, they will have a greater tendency to include UWB in their offerings. If ease of use is a significant factor in the decision and the device communicates principally with a PC, there is a good chance that they will employ a wireless USB flavour of UWB in the design. Also, if the integrators are concerned about potential spectral-saturation issues, it is increasingly likely that they will employ UWB in their design.

In a perfect world, it is not believed that WiFi and UWB materially overlap in terms of the applications they address. But the world is not perfect and a lot of factors will play into an integrator's selection of which and how many radios are employed. Too many variables are present for this facet of the market to be projected with any degree of assurance. What can be said most reliably is that both technologies will be deployed and will ultimately find equilibrium. But this will take time.

9.3 Summary

In the future of UWB, there will be a combination of changes that are already expected. Some changes will be the result of planned development by designers. Others will be the result of broader industry forces with which UWB manufacturers must cooperate. Under the heading of planned changes, UWB developers are pushing the technology into two principal directions: faster speeds and lower power consumption.

Faster speeds are designed to reduce the amount of time that a consumer must wait when attempting to download a large file, such as a movie or an audio library. Faster speeds are also needed to support streaming-video applications, such as wireless monitors or displays.

Lower power consumption is part of an effort to accommodate mobile, battery-powered devices. In small devices, it is frequently more important to use the battery sparingly and extend the battery life than it is to transfer data at the maximum possible rate. In most cases, the internal limitations of these small devices ultimately act as a bottleneck for high-bandwidth transfers.

Under the category of broader industry changes, there are two that are apparent today. Multiple radios are being integrated together into a smaller number of chips to help support greater capability on smaller devices such a mobile phones. The second major trend is the blending of PAN, LAN and WAN networks. There is a need to start exchanges on any one of these networks and to gradually migrate to another. This will require modifications in each area to first make these transitions possible and then to make them invisible to the consumer.

10 Regulating ultra-wideband

After more than 80 years of using a regulatory environment that is characterized principally by frequency and, to a lesser extent, spatial division to avoid interference, UWB introduces the significantly different concepts of spectrum underlay and 'detect and avoid' (DAA). Both of these concepts are somewhat experimental and clearly evolving over time. It is reasonable to expect that regulations governing UWB around the world may change somewhat over the next few years as we learn more about what does and does not work well in the UWB experiment.

Additionally, regulators are looking to see which applications the industry will choose to deploy UWB for. Some applications are potentially problematic from an interference perspective and others are relatively inert. The choices that UWB manufacturers and implementers make about applications will have an effect on the regulatory environment. If the industry elects to deploy a greater number of problematic applications than was estimated during the regulatory hearings, this will be of material interest to regulators and may cause them to introduce regulatory limits to discourage that deployment. The conscientiousness that the UWB industry shows in its efforts to avoid interference to incumbents will also influence how generous regulators are toward UWB interests in later rounds. As an example, if the UWB industry pushes uncoded and minimally coded video heavily in the crowded lower frequencies, it may be necessary for the regulators to issue more stringent rules to protect the incumbent services. This application is known to cause significantly more interference than something like backing up a hard drive. The UWB industry could possibly choose to deploy only coded video or the industry collectively could choose to cap the amount of bandwidth that is available to video applications. If the UWB industry uses one of these techniques to behave cooperatively, it is possible that additional frequencies might be allocated in a spectrum that was previously thought too sensitive to allow experimentation. The manner in which the UWB industry elects to deploy UWB will go a long way to determine how the regulators view this experiment that they have initiated. For these reasons, your understanding of the regulatory environment is important both to your product's viability and the future viability of the industry.

10.1 An overview on regulation

Understanding the current regulatory discussion on UWB requires some grasp of the past regulatory models so that it is easier to understand why UWB is different. The following discussion is intentionally superficial and is intended only for the purpose of giving the reader a feel for the environment rather than a detailed theoretical treatise on spectrum regulation.

For about the last 80 years or so, the method of allocating spectrum has remained essentially the same. Spectrum is divided into frequency blocks or allocations and then given (or auctioned) to various corporate and government interests to use. When administrations started to use this type of model, they found that there were opportunities to improve its efficiency. For instance, it makes no sense to allocate an FM radio frequency nationwide to a single radio station if that radio station only operates in a very limited geography. It is a more efficient use of the limited spectrum to allow other radio stations to use the same frequency, so long as they remain sufficiently separated. Likewise, it is also more efficient to allow two services to occupy the same spectrum, so long as regulatory authorities believe that the manner in which they will be deployed and used will prevent (or minimize) them from interfering with each other.

To address these opportunities for improved efficiency, administrations employed licensing. Licensing provided legal permission to use spectrum and established the terms under which it could be employed. Some license terms limited the geographical deployment, some limited the applications for which an allocation could be used, and some license terms established a priority between users who shared an allocation. These license terms conveyed to the user something very much like rights of ownership to the spectrum (although it may not be exclusive and it may be under very limited terms of use). This was the environment that existed when manufacturers wishing to build UWB radios first approached the FCC to request spectrum allocations.

10.2 The beginnings of UWB regulation

In 1998, the FCC opened the first hearings on UWB. [1] At that time, some UWB proponents were making outlandish claims about the capability of UWB to work across all frequencies, from DC to that of light, at any power level without any interference to incumbent services. Some of the claims that were made at this time would have made any snake oil salesman proud. Not surprisingly, the FCC found these claims to be invalid. Ultra-wideband was simply another form of radio, which had the potential to interfere and which operated on well understood physical principals. But, in reviewing UWB, the FCC did find an opportunity to introduce a very short-range radio design with a novel approach to spectral reuse.

The FCC came up with a concept referred to as a spectrum underlay. In reviewing the way in which the majority of radio systems were deployed, it became clear that these incumbent devices tended to generate a (relatively) high-power energy signal when they were active. It also became clear, by taking spectrum-analyzer readings in urban areas, that most spectrum licensees were not employing their spectrum allocation fully. At any given time, much of the spectrum was either not used or under utilized.

The FCC elected to create an allocation, which allowed UWB to operate between 3.1 and 10.6 GHz, an incredibly large amount of spectrum, but at very low power levels. These limits were selected to satisfy several requirements. First, the 3.1 GHz limit was established to make sure that UWB did not have the potential to affect a number of very critical services operating in the crowded lower part of the spectrum. As UWB is a regulatory experiment of sorts, this conservatism is appropriate and necessary. The second requirement being satisfied was that the bandwidth needed to be very large to offset the very low power levels and to enable systems to provide high-data-rate services. The upper level of 10.6 GHz was selected somewhat arbitrarily as a stopping point. It was viewed that, at this level, there was sufficient bandwidth available to UWB. Given that it was unlikely that technology would be able to use higher frequencies for the near future, there was no compelling reason to consider allocation of additional frequency at the time.

The power levels were set equal to those already allowed for unintentional emissions (known as Part 15 levels). The assumption was that if both devices were simultaneously active, the low power of UWB would cause no greater interference than that caused by other equipment already present in the environment and should therefore not cause additional interference beyond levels that the incumbent system was already designed to handle.

Additionally, the incumbent system would overpower the UWB device and force the UWB device to avoid the stronger signal in order to operate. To compensate for these limitations, the UWB devices were given a very large spectrum assignment. If the UWB manufacturers could adapt to these conditions, it should be possible to insert additional users into an already crowded spectrum without harming the existing occupants. The FCC viewed this underlay approach as a form of 'free spectrum'. They could enable a new industry to operate without creating a separate allocation and without causing harmful interference to existing incumbent services.

10.3 Protection vs. innovation

The FCC's decision to allow UWB was like a starter's pistol, which began a worldwide discussion on the relative merits of UWB and the concept of an underlay strategy to increase the efficiency of spectral use. But, in the rest of the world, the complexion of the discussion was significantly different. In the USA, the regulatory process is structured in a manner similar to a judicial hearing. The FCC proposes a rule and then opens debate. People are encouraged to comment in favour of the ruling and against it. Each side presents studies and arguments supporting their view of the situation. In the end, the FCC weighs these arguments and decides the matter.

A second characteristic, while not unique to the USA, is not universal to regulatory bodies. The FCC has a split responsibility. On one hand, they are responsible to protect incumbent services from interference so that they might enjoy the benefits of the spectrum to which the license entitles them. On the other hand, the FCC is responsible for promoting technical innovation to make increasingly more efficient or more valuable use of the spectrum. In some countries, there is an exclusive obligation to protect the incumbents. Countries that have a sole obligation to protect existing users tend to be much more conservative when they engage in international fora that are designed to achieve spectrum harmonization, such as the International Telecommunications Union (ITU). This difference in regulatory philosophy is the root of much disagreement that occurred internationally.

To make matters just a little more challenging, much of the rest of the world operates on a consensus model, as opposed to the FCC's judicial model of decision making. Rules are established based upon mutual agreement. This type of approach is an absolute necessity for places such as Europe, where a large number of countries exist in close proximity. Greater caution must be observed whenever a regulatory decision has the potential to generate interference in neighbouring countries. In Europe, the number of countries present makes it very difficult to solve the problem by direct negotiation between two countries at a time (a tactic that is more plausible in North America). It is much more efficient to gain agreement collectively. Additionally, if the class of devices being considered is highly mobile and the country borders are permeable, as in the European Community, international agreement on spectrum use is an absolute necessity. Unfortunately, a consensus approach involving countries with disparate regulatory philosophies can be a very challenging environment in which to introduce new concepts.

10.4 European regulatory leadership

From this point on, the conversation will be focused on the work in the regulatory bodies of the European Community. In practice, the USA frequently acts as the lead rabbit or innovator in terms of identifying and promoting new technology. The FCC's resources, permissiveness toward innovation and organizational structure allow it to move to a conclusion quickly. In the case of UWB, the FCC decision preceded a European decision by almost five years. In contrast, Europe leads the world in terms of consensus deployment. This is visible in the ITU proceedings. In the ITU, decisions are reached by consensus. The USA expresses a single opinion (not a vote) in committee work, which regularly contains several dozen participating administrations. When European administrations enter the worldwide ITU proceedings, they have already established Europe-wide positions and are more inclined to speak as a collective. This represents a block of approximately a dozen or more administrations supporting a compromise position.

This dynamic allows the European position to have an asymmetric influence on the world direction in regulatory affairs. Non-aligned Asian, African and South American administrations are more likely to feel comfortable with the European consensus posture in most cases. The nature of the European consensus approach vs. the FCC judicial approach to decision making will also tend to cause European decisions, and likewise ITU decisions, to be more conservative than USA positions. Because of Europe's influence on worldwide regulatory decisions, a deeper discussion of the European regulatory process, as well as its proceedings on UWB, is warranted.

10.5 European regulatory bodies and organization

As a beginning statement, the reader should recognize that the present structure in Europe was designed in the manner of an old house. There was an original structure, which functioned for a limited purpose. Over time, additional demands were placed upon it, requiring that it be extended and renovated significantly. The final design, while possibly less than elegant, achieves its intended purpose. In describing the European institutions, it is useful to follow the same path in explaining the system that they took, over time, in building it. The final product is much more understandable when taken in this manner.

10.5.1 The national administration

When European spectrum regulation was first established, each country worked almost autonomously. Spectrum was (and is) a national asset, which is under the control and ownership of a national administration. There are some exceptions for cases when emissions cross borders, but these are outside the scope of this conversation. At the very base of the European system of standards and regulations is the national administration. Each administration is responsible for establishing the laws governing its populace. It is the only group which has this authority. Each administration has rules governing its theory of how regulation should be pursued. In the UK, they have a policy of balancing protection of incumbents with the development of technology in a manner similar to the USA. In France, they have a policy that increases the weight given to protection. These administration-specific rules and philosophies are the background on which the rest of the picture is painted.

10.5.2 CEPT

The next level of the puzzle occurs when countries attempt to cooperate with each other. This occurs for a variety of reasons. Some services, such as very sensitive radio astronomy, will be affected by emitters that may be in another administration's jurisdiction. These situations require coordination between the affected countries to allow the radio astronomy service to operate efficiently. Some issues for cooperation are economic in nature. Imagine the time, paperwork and engineering resources required for a manufacturer to manage the sales of a single radio design in every European country. Each country may have a unique spectral allocation or license condition, which a manufacturer needs to understand and hopefully influence. Each country has unique testing requirements to make sure the manufacturer is in compliance with its rules. All of this creates a morass of red tape, which limits the ability of European manufacturers to cultivate a broader market. Because of these problems, CEPT (the European Conference of Postal and Telecommunications Administrations or Conférence Européenne des Administrations des Postes et de Télécommunications) was created.

CEPT is a body that is created by international treaty for the purpose of harmonizing work in telecommunications among the member countries. It is a voluntary organization. If a CEPT decision is not acceptable to a member administration, that member simply declines to sign the decision and it is not bound to comply with the decision's findings. If an administration chooses to sign a CEPT decision, it has an obligation to enact the appropriate national laws to make the CEPT decision apply to the nation's citizens.

10.5.3 The European Union, European Commission and Radio Spectrum Committee

While the CEPT structure was clearly a substantial step forward in that Europe would now have a mechanism to coordinate its activities to some degree, there was a perceived need for Europe to compete against the USA and various developing economies. These groups were perceived as being more economically efficient markets because a single decision was applicable to the entire market area. A single currency was used universally throughout the market and goods and services were able to flow readily across the region. On the basis of this need, the European Union (EU) was created for the purpose of removing barriers, which inhibited the effective growth of commerce.

As part of the creation of the European Union, the European Commission (EC) was created to be the executive body tasked with administration. A committee was also established within the EC, under the name of the Radio Spectrum Committee (RSC). The membership of the RSC was composed of national administrations. The membership structure was established to make sure that the policies of the EC effectively represented the interests of the member states.

In spectrum regulation, the creation of the EU and EC causes some noteworthy changes to the spectrum-regulation environment. The first change is that the EC has been vested with authority to initiate technical investigation by CEPT and the technical committees of the ECC

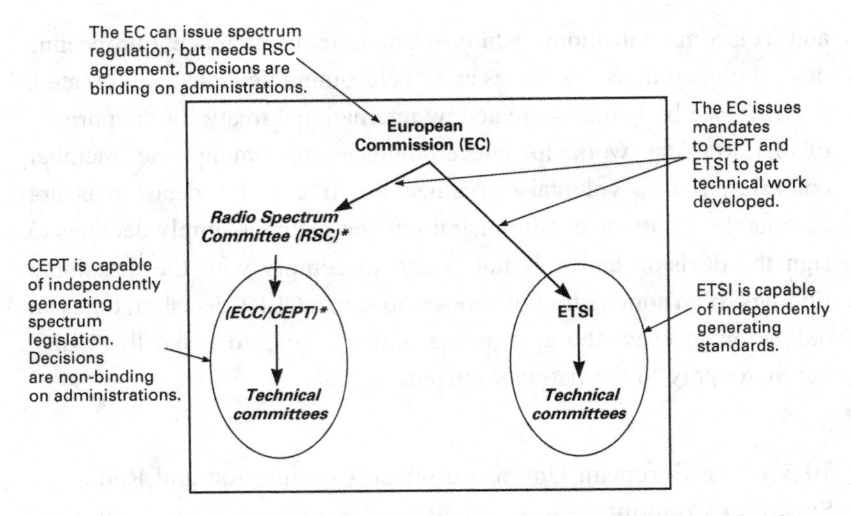

Figure 10.1 European institutions

(Electronic Communications Committee) through the use of a mandate. Mandates can also be used by the EC to direct ETSI to develop standards related to a new technology. The EC was given this authority to enable it to be proactive in the development of the EU's economy. Figure 10.1 attempts to capture these relationships.

The relationship between CEPT and the EC is a little peculiar. The creation of the EC regulatory powers did not cause CEPT to become obsolete. Regulations are still generated by CEPT under the treaties that existed prior to the EU. But, layered onto the CEPT structure, it is also possible for the EC to issue spectrum regulations. The principal difference here is that while CEPT decisions are voluntary, based on an administration's agreement, EC decisions are mandatory for all EU members. Each of the countries participating in the EU has an obligation to convert the EC decisions into local legislation. CEPT decisions are only binding upon an administration if the administration agrees to be bound.

But, to add one more twist to the puzzle, the treaties establishing the EC tie the EC's spectrum authority back to the member administrations in the form of the Radio Spectrum Committee (RSC). This is to say that the treaties allow the EC to begin investigations and issue mandates beginning technical evaluation and standards development of its own volition, but it must have a critical amount of national administration support in the form of RSC decisions in order to authorize a binding EC decision on spectrum policy. The law enabling the EC to issue spectrum rules requires that the EC do so with member support expressed through the decisions of the RSC.

Using this power, the EC mandated CEPT to generate regulations to enable the use of UWB in Europe. CEPT, working in cooperation with the European Communications Commission (a body tasked with engineering study) then assigned Task Group 3 (TG3) to perform the analysis and to recommend the regulatory structure needed to make UWB occur.

It does not take keen observation to realize that the regulatory procedure in Europe can be a bit challenging to follow for anybody who has not taken vows of poverty and chastity and spent several years in silent meditation on the topic. The above description does not come close to doing it justice, but it may be sufficient background to explain some of the interplay going on today and in the near-term future.

10.6 The challenge of regulating UWB in Europe

Contrary to the opinion of some of the participants, the contests occurring in the regulatory process are not between the forces of good and evil. While one might wish for such a black-and-white understanding of the resulting rules, in practice, the process explores an exhaustive palette of grey. Regulators are being asked to make decisions that position the interests of incumbent organizations who are using the spectrum today and the value to the economy (and incumbent organizations) that they provide against prospective users of the spectrum and the potential benefit that may be gained by the economy if they are permitted to use the spectrum for new applications. In practice, this frequently means finding a compromise where both sides are equally unhappy, but are able to function.

In the case of UWB, the challenge has been made substantially greater than usual because the spectrum under consideration spans such a large frequency range (3.1–10.6 GHz in the United States) and overlaps so many incumbent services. Adding to the complexity of the problem is the fact that UWB devices are expected to be very small, very cheap, highly mobile and intended for consumer use. This means that the technology is not readily amenable to licensing constraints, which might otherwise be used by regulators to limit a new radio technology. As a consumer product, regulators are forced to assume that it can go anywhere that a consumer can go and might be used at any time. To add a final degree of complication to an already challenging condition, the exact nature of the applications for which UWB will be employed is uncertain. While industry participants advocating the use of UWB understand the initial applications for which it will be employed, it is the nature of technology to evolve over time and new applications are likely to emerge.

The problem of regulating UWB is somewhat moderated by the very low emitted power levels being considered. Because UWB emits approximately the same amount of power intentionally that many electrical devices, such as computers and televisions, emit as a by-product of the operation of electronic components, UWB emissions tend to be fairly well contained or at least attenuated by building structures, automobile bodies and the like. Ultra-wideband's power limitations tend to constrain interference questions principally to those incumbent services operating in close proximity to a UWB transmitter in terms of frequency and physical separation.

10.7 The first mandate – technical work begins

Regardless of the difficulty of the challenge, the European Commission first undertook an effort to enable UWB in February 2004. At that time, the EC issued the first mandate, which requires the ECC to create appropriate regulations. As part of the same mandate, ETSI is directed to create the necessary standards to allow UWB to be legalized. The work in ETSI has largely been concentrated upon creating measurement standards to ensure compliance of UWB devices to regulatory requirements. At the time of this writing, the ETSI work has yet to be completed. No further discussion on ETSI standardization is offered at this time. The portion which follows discusses the creation of European regulations for UWB. This narrative is being presented in a sequential form to simplify the discussion for the reader. In fact, there was a great deal of parallelism that was employed in the development of various documents. This parallelism is responsible for a number of the inconsistencies that will be evident to somebody reading through the resulting paperwork.

10.7.1 Characterizing ultra-wideband

In mapping out the technical debate that occurred in ECC/Technical Group 3 (TG3), the first problem for TG3 was to characterize UWB's behaviour. To estimate interference on incumbent services, it is necessary to understand the proposed service that may be generating the interference. One must understand the details about how the radios behave and will be used. For instance, devices operating at low power and high frequency indoors would be expected to cause interference to fewer incumbents than would be generated if the devices would be used outdoors, at high power or at low frequency.

As a first step, UWB's frequency and power-emission characteristics were taken from the US regulations for UWB for the purpose of creating a starting point for the analysis. The USA was the only administration who had issued rules at this time. The majority of manufacturers who were designing UWB devices were doing so based upon the US limitations. But, power limits and frequency ranges only described how a UWB device would behave when it was switched on.

Since it is fairly evident that devices such as cameras, computers and mobile handsets would not be switched on continuously, it was also necessary to perform additional analysis about the specific applications in which UWB would be employed to characterize the frequency with which they would be on the air. To do this, an activity analysis was performed in TG3 to describe anticipated customer applications. For each application, data rates, typical data-transfer size and duty cycle, range and activity factors were determined.

Two principal conclusions emerged from the analysis. First, UWB interference is overwhelmingly dominated by the use of video. In fact, video accounts for approximately 95% of the interference potential of UWB. The second major fact is that UWB will behave principally as an indoor technology. Both of these facts are largely attributable to the same line of logic. Because UWB is very fast and will get faster over time, it tends to get on and off the air very quickly for any transfer of a bounded length. Ultra-wideband's speed minimizes the interference potential of applications involving file transfer. This includes transfer of music files, video files, print and scan functions as well as the vast bulk of Internet communications.

Where UWB's speed is less of a factor is in streaming services. Streaming services, such as broadcast video, require a high volume of data to be transferred every second. As an example, an MPEG2-encoded high-definition video stream requires up to 24 Mb/s for as long as the video is on. Following this logic a little further, the only place that users are expected to have access to constant video streams is from broadcast-video infrastructure, which is principally present (and overwhelmingly consumed) in the home.

Creation of a broadcast infrastructure using UWB devices that would not be indoors may well be physically possible, but is generally thought to be implausible. For a broadcast-video provider to establish a UWB infrastructure outdoors would require that they establish a grid of UWB transmitters approximately every 10 m to enable the consumer to remain in constant contact with the streaming infrastructure. Furthermore, very simple restrictions, such as a prohibition on fixed outdoor antennae (employed in the USA), quite readily precludes such designs from being developed even on a small scale. Because of these points, it was concluded by TG3 that UWB could be treated overwhelmingly as operating indoors.

10.7.2 Evaluating UWB's interference potential

Once the activity factor analysis was completed and the UWB RF characteristics were established, the incumbent services were then able to determine whether they would be affected by UWB emissions. Each service drew up scenarios describing potentially harmful interference, which could then be debated in committee. Given the extremely wide frequency range over which UWB could potentially operate, the number of incumbent services involved was substantially larger than most previous proceedings. A short list of the affected services includes cellular telephony, point-to-point microwave services, wireless broadband, (civilian and military), radio astronomy, aviation (navigation, altimeters, communications, etc.) and a variety of others. Not only were existing services considered, but future services, such as 4G wireless telephony, were also debated.

For each of these incumbent services, it was necessary to describe the incumbent service to some degree (and to a lesser degree for military services maintaining secrecy) and then draw up a scenario under which harmful interference would occur. Task Group or one of its subcommittees would then debate the assumptions built into these scenarios and attempt to reach the consensus that the protection measures required.

The interference scenarios could broadly be characterized in one of two classes. Single-entry interference scenarios were drawn when the incumbent services believed that a single UWB device had the potential to create harmful interference. In single-interferer cases, the propagation calculations were fairly straightforward. Normally, a line of sight existed between the UWB radio and the incumbent antenna. So, debate shifted to other issues:

- How much interference is an incumbent required to tolerate?
- How is the interference budget into a given band fairly shared between several sources of interference?
- Should the calculations be run with an assumption of no existing interference, such as self-inteference produced by other cells within a network, noise present through unintentional emissions, etc? If not, what level of existing noise is it reasonable to assume?
- Should harmful interference be measured in terms of the communications impact on a single user or the business impact on a whole network?

Many of the issues that this discussion brought up were beyond TG3's ability to resolve for all practical purposes. They frequently involved broader policies, which had yet to be set or clarified. As an example, establishing a budget for all possible interference sources that is valid under all conditions at all times is very difficult at the least and may well be impossible. It certainly is not present today and is well beyond the ability of TG3 to create. But, from an incumbent's perspective, this is still a perfectly reasonable question to ask. A point surely exists at which an unacceptable amount of interference is introduced into a band in which the incumbent has a license, which supposedly protects them from this issue. At that point, some relief is justified.

The second major class of analysis involved aggregate interference. In this scenario, a single UWB device is insufficient to cause harmful interference to an incumbent service. However, if one turns on 1 million UWB devices simultaneously and places them all within a single floor of a building, the noise floor may potentially rise to a point that causes harmful interference for the incumbent service.

In calculating the aggregate interference, one must establish how many UWB devices are active (from the activity analysis) and then allocate this number between indoor and outdoor locations. Outdoor locations would automatically be assumed to have a line-of-sight path loss, while indoor uses would have additional losses due to building structures.

Even the seemingly straightforward discussion about path-loss characteristics for indoor or outdoor uses was fertile ground for debate in TG3. Devices operating in close proximity to a window would potentially have path-loss characteristics that were almost line of sight. Buildings that were largely open except for cubical walls would clearly experience significantly less attenuation than would buildings with solid-wall construction. In office complexes and metropolitan areas, it was conceivable that buildings with lower path-loss characteristics might begin to dominate the interference profile. Task Group 3 had to discuss all of these situations.

Unlike a satellite application, where one might reasonably argue that the antenna aperture captures a sufficiently large population that the effects can be averaged, point-to-point systems located in an urban area are not so easily solved by averaging. The effects of a reduced path-loss exponent in an urban environment are a very real possibility and need to be taken into consideration separately. In this case, TG3 members conducted a number of simulations to try to characterize the interference potential and to select path-loss exponents that more appropriately characterized the interaction between aggregated UWB interference and the incumbent system.

10.7.3 A zero-interference assumption

In both the single entry analysis and the aggregate interference analysis, a common assumption persisted. Task Group 3's recommendation would be made at a point where zero interference was created into incumbent services. This position reflected the most conservative administration posture designed to protect incumbents. Not surprisingly, this is also the position that most incumbents prefer. Considering that UWB was a new technology with a significant number of uncertainties about its use, it is also argued that this is the most reasonable conservative position.

The zero-interference assumption showed up in the selection of interference scenarios. If it was possible that an interference condition could exist, no matter how unlikely, it was assumed that the regulation would be crafted in a manner that precluded that possibility. The assumption showed up in the activity analysis. Activity was stated assuming very inefficient video implementations and simultaneously ignoring proven innovations that would reduce video activity (such as MPEG 4, which was well deployed at that time). With this philosophy in mind, TG3 made its recommendations on the regulation in the form of Report 64.

10.7.4 Report 64

Once Report 64 was approved in TG3, it was advanced through CEPT and the ECC to the RSC and EC as a response to the EC's first mandate. At this point, it is useful to note that the relationship between TG3 and the EC in this proceeding is consultative. Task Group 3 acts in the role of a consulting engineer, whose data are intended to inform the EC. The EC was not required to accept the conclusions that TG3 offered.

Upon receipt of Report 64, the Radio Spectrum Committee responded with the following statement (a portion of which is excerpted here):

> To recall, CEPT was mandated to undertake all necessary work to identify the most appropriate technical and operational criteria for the *harmonised introduction* of UWB-based applications in the European Union. The Commission's current assessment of the activities in this respect is that the underlying objectives of the Mandate would not be met if the CEPT work is finalised exclusively according to the technical results presented in the Interim Report
>
> It is widely accepted that the conditions of the use indicated in the Interim Report would satisfy the essential requirement of protecting existing radio services from UWB-generated harmful interference. The Commission fully understands and supports the need for critical, often economically valuable sectors to minimise the potential negative impact of UWB on their activities...
>
> However, the likely consequences of having EU-wide or national regulation based on the results of this Interim Report would be to delay sine die the regulated development of a mass market for UWB-enabled devices in Europe, thus undermining the potential benefits associated to the rapid uptake of this technology. Furthermore, in the absence of reasonably regulated access to UWB devices, there would be a clear risk of an uncontrolled proliferation of illegal equipment across borders into Europe, and of a de facto acceptance of the more liberal FCC rules.

Figure 10.2 is a summary diagram, which captures the concluding work from Report 64. The top line represents the power permitted by the FCC regulations in the USA, while the lower line was the power recommended by the TG3 committee. The substantial difference between the two is principally a result of the zero-interference logic used by the committee.

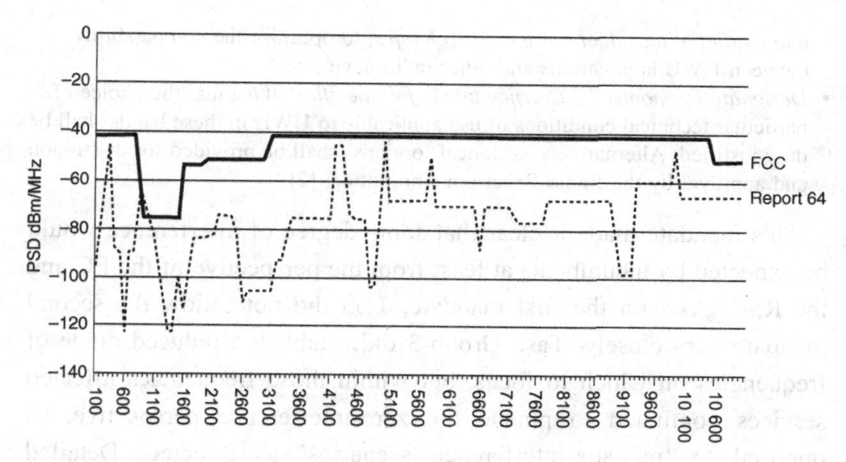

Figure 10.2 Report 64 spectrum mask vs. FCC

10.8 The second mandate

At this point, the European Commission then issued the second mandate, in pursuit of an alternative strategy to gain a less conservative and less theoretically based recommendation. Rather than asking CEPT/ ECC and TG3 to redo work that had required several years to accomplish and involved a great deal of debate, they elected to search for an alternative to the power mask recommendations in Report 64. The second mandate required that TG3:

- *Adequately schedule and prioritise activities* under this mandate to optimise the possibilities for a successful conclusion of the work in a timely manner and to reflect activities already undertaken in this area;
- *Determine the frequency range(s) to focus upon initially* for specific UWB applications, and justify this selection on the basis of clear criteria (such as maturity of products in such bands); study the possible use of additional frequency ranges in the future;
- *Undertake complementary technical studies* between UWB applications and potentially affected selected radio services, notably for the priority frequency ranges, based on realistic interference scenarios;
- *Report on the empirical evidence* gathered by current measurement campaigns within and outside Europe. Apply such results to validate or modify theoretical coexistence scenarios between UWB applications and other radio users;
- *Carry out a detailed impact analysis on the selected bands*, for a restricted set of alternative regulatory solutions; Analyse in sufficient depth the feasibility and impact of *generic and dedicated regulatory measures, operational conditions*

and available technical mitigation techniqes, to optimise the compatability between UWB applications and other radio devices;...

- *Designate harmonised frequency bands for specific UWB uses;* the choice of particular technical conditions of use applicable to UWB in these bands shall be duly justified. Alternatively, technical 'options' shall be provided for discussion and approval by the Radio Spectrum Committee; [2]

This mandate made it clear that some degree of interference would be expected by incumbents at least from the perspective of the EC and the RSC. As with the first mandate, TG3 did not follow the second mandate very closely. Task Group 3 did establish a reduced range of frequencies on which to focus, but within those frequencies affected services continued to pursue the zero-interference perspective, as opposed to 'realistic-interference scenarios' as directed. Detailed impact analysis was not conducted. And finally, a number of use restrictions were included into the final recommendation which were not backed up by analysis (i.e., use on aircraft, rail, automobiles and in model aircraft). Despite this, a great deal of productive work was done during the second mandate effort. This was particularly true of fixed wireless-access services, the phased approach, measurement campaigns (for radar) and economic analysis.

In working on fixed wireless access (FWA) services, a great deal of work was conducted in an effort to simulate the effects of UWB on services such as WiMax. Fixed wireless-access services are most likely to be affected by the single-entry scenario. This is to say that these services expect to have their radios closely associated with PCs and consumer electronics equipment and so would be in very close proximity to UWB radios addressing the same platforms. Because of this close proximity, interference was likely.

The questions that were not well understood were the severity of the expected interference and the degree to which it could be reduced or eliminated through the use of various mitigation techniques. In evaluating the interference impact, it was debated whether the interference was limited to a single user who would not be able to detect the condition or whether the entire FWA network would be forced to compensate for interference occurring simultaneously with a number of users and thereby reduce network capacity materially.

The second block of discussion involved the use of a phased approach, wherein UWB would be allowed to use lower-frequency bands for a period of time without mitigation techniques. After the phased-approach time period expired, manufacturers would either have to move into spectrum above 6 GHz or implement methods to detect and avoid (DAA) incumbent signals when present. This avenue was investigated for a couple of reasons. Ultra-wideband manufacturers were on the verge of shipping products into the USA, Japan and Korea. If Europe placed requirements on UWB products that caused them to be delayed, it was perceived that a grey market could develop. In a grey market, devices developed for other countries would be illegally shipped into Europe and sold. The growing presence of Internet sales and the low cost of UWB devices made it highly probable that a grey market would occur. By allowing a phased approach, it became possible to sell existing designs with small modifications into Europe immediately and so eliminate the incentive to import radios.

The third block of discussion that took place during the second mandate centred on efforts to quantify the interaction with radar and radio-astronomy services (RAS). To resolve these questions, there were a number of additional studies conducted. Several radar measurement campaigns were undertaken to attempt to quantify the impact on various radar systems when exposed to prototype UWB devices or signal generators that simulated UWB emissions. Aeronautical and military radar operators were concerned that if UWB devices were operated close to radar, the antennae would reduce the sensitivity of the systems and make them less effective.

In radio astronomy, additional analysis was performed by RAS groups to try to quantify the protection zones required around RAS ground installations. Radio astronomy services were concerned that a small number of devices in close proximity to the RAS antennae would be able to interfere with the very-low-power signals being measured. If this were true, exclusion zones would need to be established to keep UWB devices at a distance, such that interference was managed.

In addition to the technical work, the second mandate required that an economic impact assessment be performed to assess the effects of the proposed power mask as well as the phased approach. This was intended to put a monetary value on the trade-offs being made. Clearly, if the introduction of UWB were to have no better than a neutral to negative economic impact on the European Community, regulators would have a strong disincentive to make concessions aimed at enabling UWB's introduction. Conversely, if UWB were to have a significant positive economic impact, it would be more likely to warrant taking pains to make introduction occur more quickly and with fewer obstacles. In the end, economic contributions were submitted to the ECC from two sources.

Ultra-wideband proponents quantified sales expectations reflecting a positive effect of several billion euros. The second contribution, which is known as the Mason Report, was underwritten by the UK regulatory agency (Ofcom). [3] The Mason Report had originally been commissioned by Ofcom very early in the regulatory processes and so was not a direct response to the second mandate. Because of this, some of the analysis was based upon data that were no longer current. On the positive side, the Mason Report made an extensive effort to quantify the negative impact on incumbent services. The Mason Report projected a multi-billion-euro gain to the UK from the introduction of UWB.

Work on the second mandate concluded at the end of 2006. A final report was submitted by TG3 and accepted by both the CEPT/ECC and the EC. At this point, the EC felt that it had sufficient agreement between administrations to support an initial ruling. In March 2007, the first European decision on UWB was issued. [4] The decision enables unlicensed UWB devices to be sold in all countries who are members of the European Union. The EC also issued its third mandate to clean up several outstanding issues.

10.9 The third mandate

The TG3/ECC recommendations that came out of the second mandate included several provisions for which analysis had not been conducted. Specifically, the recommendation proposed excluding the use of UWB on road or rail vehicles, and in cars and aeroplanes. Application-specific exclusions of this type are challenging to implement in a European legal context. The definition of a road or rail vehicle must be precisely defined in order for the rule to be enforced. The rationale for discriminating against specific applications must be clearly justified.

The recommended exclusion of UWB from aeroplanes also had certain challenges. Rules governing the operation of transmitters on board aircraft are usually managed by aviation authorities, who have legal authority to set rules for radios which it is allowed or prohibited to operate on board aircraft. Task Group 3 would not normally perform investigation into these areas, as this would lead to overlapping and potentially contradictory legislation. To justify TG3's investigation into aircraft uses, it would need to be argued that aircraft uses of UWB are likely to cause interference with ground-based radio systems. Based upon the lack of analysis conducted by TG3 on this topic, it was unclear that any foundation existed for the recommended prohibition of UWB on board aircraft. As part of the third mandate, TG3 was asked to review their findings on these topic areas.

Prohibiting the use of UWB in automobiles was similarly problematic. At the conclusion of the second mandate, no work had been conducted by TG3 that would support the conclusion stated in the recommendation. The third mandate was intended to correct for this deficiency. Measurement campaigns were conducted on cars during the third mandate, to determine the attenuation that is provided by the car body. If the car body attenuated emissions at a rate similar to the levels projected for a house or structure, the use of UWB within an automobile could reasonably be thought of as just another indoor use. In this situation, it would be inappropriate to attempt to preclude the use of UWB in cars. If the attenuation of the car body were to be found to be much less than a building, then additional investigation would be required to characterize the volume of activity in cars and its potential effect on affected services.

Also as part of the third mandate work, additional review was requested by the EC into the upper boundary of the allocation. The TG3 recommendation from the second mandate advised that the upper limit of the allocated band be carved back from 10.6 GHz to 9.5 GHz. This change was justified at the time from radar studies, which had been conducted in lower bands and which were extrapolated to higher frequencies. In committee discussion, it was unclear that the extrapolation was properly conducted and interpreted. As part of the allocation that was granted to UWB, the EC elected to take the more conservative position and allow the carve-back, but also elected to conduct additional investigation in the third mandate. In this move, the EC was trying to encourage the harmonization of spectrum use around the world, unless there was a clear reason not to do so.

10.10 Single entry vs. aggregation

Now that the institutions and the processes are reasonably well described, it is time to talk about the classes of interference that are being considered by regulators. Taken very broadly, interference is evaluated on the basis of single-entry effects and aggregation. In single-entry evaluation, it is assumed that the emissions from a single UWB radio are sufficient to cause harmful interference to a protected service. Given the very low power that is used for most UWB radios and all HDR and LDR radios, the services likely to be affected are usually those that are operational in extremely close proximity or those with very high-gain antennae, which are likely to capture a UWB transmitter in or near the aperture.

WiMax and other fixed wireless-access services were evaluated for single-entry effects, where the service is in close proximity to the UWB radio. Given that both UWB and WiMax would probably be present in and around PCs and in other platforms likely to collect multiple radios, the probability that they would come within interference range was fairly high.

During discussion in the regulatory committees in Europe, it was eventually agreed that if devices were present within a common system, such as a PC, it fell to the PC integrator (or to the customer buying radio cards and assembling a system) to develop methods to manage interference between radios. It was believed that radios interfering within a common enclosure were beyond the limits where regulations should be imposed. In this situation, the integrators had a significant number of tools at their disposal to avoid interference and also possessed a common interest in making the device, which was shared with both radio services, succeed. Ultimately, it was expected that the market would resolve the problem without the need of additional regulation.

If, however, the WiMax device were either completely independent of the PC, as in the case of Wimax in a mobile phone, or if Wimax were in a modem attached to the PC through a USB or serial cable but not embedded within the body of the PC, it was determined that regulatory limits were appropriate to assure protection of the fixed wireless-access services. 'Detect and avoid' is the principal tool that is employed to limit interference in these cases. It will be described in greater detail later in this chapter.

Radar systems are another single-entry case that was investigated extensively during the regulatory proceedings. Unlike WiMax and fixed wireless access, which were at risk from being within very close proximity of a UWB device, radar tended to have very high gain antennae, which would effectively amplify the UWB signal if it were to be generated anywhere in or near the path of the radar signal. While the physical separation was much greater than the fixed wireless-access case, the same interference effects would be felt. Like fixed wireless access, radar services are also expected to be protected through the development of DAA.

The final service with a demonstrable weakness to single-entry effects is radio astronomy (RAS). These systems employ extremely large parabolic dishes, which are pointed at some portion of the sky to search for a range of naturally occurring radio emissions. While it is highly improbable that a UWB device would actually find its way into the aperture of the antenna, it is possible that a device could get close enough to the very sensitive RAS system to disturb its reception.

Because RAS systems do not transmit but are instead receive-only, techniques such as DAA are not effective in protecting them. Instead, it is necessary to establish exclusion zones around recognized RAS sites. As these sites are generally located well away from urban areas to escape the ambient noise that a city generates, the exclusion zone is not a material inconvenience.

Even if a UWB device proves not to cause single-entry interference to a service, the story is not over. It is also possible that several UWB devices operating simultaneously are sufficient to cause interference where a single device could not. In general, this type of problem is of concern where antennae are focused at a large area, which may contain some number of simultaneously operating UWB devices, thus raising the noise floor over that area. It is also possible to generate an aggregation problem if several devices operating near or in the path of a receiver can have the effect of generating a rise in the noise floor.

The low-power, noise-like nature of a UWB signal made it somewhat difficult to make a case to describe UWB as causing unique harm. The proliferation of electronic devices that unintentionally emit at levels similar to UWB is already having the effect of raising the noise floor, which has necessitated that radio services design their systems to accommodate that floor. The argument that has to be made in aggregation is that UWB devices will generate conditions exceeding those for which the incumbent systems are already designed. If this argument cannot be made, no harm is present and regulators have a greater inclination to allow UWB, as it expands the economic base of the region.

With that said, there is one area where UWB has the potential to aggregate and to cause interference. This is in the case of uncompressed or minimally compressed video. If a UWB radio is transmitting files, asynchronous data or other non-streaming traffic, the data are sent very rapidly by high speed UWB systems and the radio then is quiet. The ratio of on-time to off-time is strongly weighted in favour of off. To make matters a little better (from an interference perspective), the coordinated nature of the UWB piconet prevents transmitters from attempting to emit at the same time and so the power in a local area is less likely to be combined.

In the case of uncompressed or minimally compressed video, the situation is different. Uncompressed video can readily occupy data bandwidths of 1.5 Gb/s to 11 Gb/s. This has the potential to cause a UWB radio to be on continuously. Further, if the application is a wireless display showing television programming, it is not uncommon for more than half of equipped users to be on simultaneously during prime time hours. Because of this, deeper investigation was warranted.

Ultimately, it was decided that the probability of worst-case aggregation was relatively low and easily monitored over time to see if assumptions were mistaken or changing. To begin, the business incentives to use UWB for a wireless video application were relatively low. Display manufacturers were attempting to eliminate decompression logic. As UWB is only capable of 480 Mb/s in the first generation, with a roadmap to 960 Mb/s in the next 2–3 years, compression could not be eliminated and so the financial incentive for manufacturers was eliminated. Additionally, the life-cycle of a television is approximately eight years. Maximum penetration would take a period greater than eight years to reach, giving regulators time to monitor the situation. As a third item, if the full bandwidth is used for video, it becomes impossible to put any other application in that spectrum. This eliminates all of the other PC and CE applications, which is a situation which manufacturers would not accept. The fourth issue is one of video-compression trends. A high-definition video stream today can take somewhere around 19–24 Mb/s using MPEG 2 encoding. More advanced coding techniques are already well deployed. As an example, the same stream using MPEG 4 encoding is capable of dropping that data rate to approximately 10 Mb/s. It is expected that further coding improvements are likely. And finally, industry organizations, such as WiMedia, Bluetooth and Wireless USB, have put in place maximum bandwidth and sharing rules that limit the use of video.

One of the principal services concerned about the aggregation effects of UWB was the satellite industry. Satellite distribution of data and video, as well as satellite-based earth-monitoring systems, had the potential of negative effect if it was demonstrated that UWB would significantly change the noise floor. Since both of these services were outdoors, one final factor came into play. Video is overwhelmingly an indoor activity. For interference to reach these systems, it is necessary for it to get through the building structure that materially attenuates the signal. Incumbents questioned whether UWB would in fact operate as an indoor service.

For UWB to operate as a streaming-video link and risk aggregation, it is necessary for one end to have access to a video-streaming feed. Overwhelmingly, these are indoors, coming from TV broadcasters. While it is indeed possible to have a camera feed outdoors, it is not possible to get the very high levels of consumption that one gets from the general populace watching television during prime time.

As a conservative gesture, several administrations have taken steps to further limit this condition. In the USA, a rule was issued that precluded the use of fixed wireless infrastructure outdoors. This eliminates camera surveillance and other applications, which may tend to aggregate. The Japanese required that one end of the UWB network should always be 'mains powered' or plugged in. In this way, outdoor transmissions are largely precluded.

In the European proceedings and in the ITU, it was determined that, even under the most generous assumptions, the use of UWB outdoors in any volume was extremely improbable. Even cars were evaluated for their potential to act as outdoor emission sources. Rather than issue a specific exclusion against outdoor uses, the European process elected to monitor the deployment.

10.11 The need for ongoing regulatory work

Because no UWB radios were market ready during the development of the European regulations, it was necessary to make decisions based heavily upon theory. Where possible, measurement campaigns were run, which used early prototype UWB devices or signal generators that were programmed to transmit a signal believed to be similar to a UWB device. While such measurement campaigns added a great deal to the understanding of the technology's behaviour, the testing was far from exhaustive. As an example, it was never possible to deploy a sufficient number of devices to test aggregation effects. It was also not possible to do sufficiently broad testing to thoroughly evaluate the actual path losses from indoor use of UWB into nearby fixed wireless-access devices.

The only reasonable way to deal with these uncertainties from the perspective of national administrations was to incorporate conservatism into the analysis as well as the initial rulings and then to measure the ongoing deployment of UWB devices to look for behaviour that varied from the assumptions used. If the ongoing measurements show a greater interference potential than was initially anticipated, regulators could then take appropriate steps to counter the effects before market deployment became significant. This could involve modified power limits, additional mitigation techniques or a variety of other mechanisms. Conversely, if ongoing measurements show that UWB is more inert than was anticipated by the theoretical analysis, it would be possible to reduce restrictions on UWB to some degree.

The European regulatory administrations expect to be conducting this type of ongoing measurement. Three years after the release of the initial ruling in March 2007, there is a mandatory review to occur. Less formal reviews are also likely to be conducted during the interim as well.

The ongoing monitoring of UWB is a point that is worthy of specific note to anybody planning to implement UWB devices. The decisions that product developers make have the potential to affect future changes to UWB regulations. For instance, if UWB manufacturers choose to implement uncoded or minimally coded video applications, this will raise alarms for aggregation concerns and for potential interference to FWA services. More conservative regulations would potentially be required to stem the growth of those video devices in the market. Given the fact that European decisions have an unusually strong effect on regulatory decisions around the world, a ripple effect in other countries is also a real possibility. Ultra-wideband manufacturers would be wise to implement techniques that reduce the use of bandwidth and power, whether they are required by regulatory bodies today or not.

10.12 Moving above 6 GHz

At this point in the discussion on regulations, there are a small number of topics that have proven to be confusing to the UWB community, and even to some regulators, which require some additional elaboration. These include the discussion about moving UWB above 6 GHz and mitigation techniques.

During the regulatory processes in Europe and in other parts of the world, it was advocated by a number of incumbent groups that UWB be given a spectrum allocation above 6 GHz only. (As a matter of fact, the Bluetooth Special Interest Group, which represents the interests of the mobile handset community in PAN matters, has chosen to delay deployment of UWB until it is possible to do so in the spectrum above 6 GHz.) A number of arguments were used to justify this suggestion. First, lower-frequency spectrum should be reserved for those services that need to be able to penetrate structures or to cover large distances. Since UWB is intended to operate as a one-room technology, with a range of 10 m or less, it could conceivably operate at a higher frequency.

In theory, this is a reasonable way to organize the spectrum use. However, there are practical problems with this logic. At the time of this writing, it is not possible to produce a radio above 6 GHz in silicon at price points that consumers can realistically afford for the envisioned uses. It is expected that advances in silicon lithography processes will make this a viable alternative in the first two quarters of 2008. Delaying a European decision until that time was possible, but only at the risk of creating a grey market. If US-compliant devices would be available in the market for 18 or more months without a European product, there would be a strong incentive for entrepreneurs and consumers to import US devices into Europe. Regulators believed that the grey market threat was significant, based upon other historical examples.

The argument for allocating spectrum above 6 GHz only to UWB also ignored the fact that attenuation effects above 6 GHz would reduce the effective range and throughput by one half that which was available in the 3.1–5.0 GHz bands. This means that instead of having a 10 m PAN technology, UWB would become a 5 m design. Under this scenario, UWB would also be significantly less able to penetrate solid objects, such as shelves and cabinet doors. This proves to be problematic for applications such as computer peripherals and consumer electronics devices. In these two classes of applications, it is not uncommon for consumers to enclose consumer electronic equipment into cabinets or to place a PC inside a desk for aesthetic reasons. This is not the case for applications which Bluetooth envisions. Bluetooth devices are anticipated to operate at a range of less than 1 m in peer-to-peer applications and at significantly lower than peak throughput rates. Because of the limitations of battery power, bus structures on handheld devices normally operate at much lower data rates than bus structures on AC-powered devices. The peak rates of UWB significantly exceed the bus rates of portable devices. Therefore, designers of handheld telephony devices are willing to trade off some of this excess throughput. So, many of the limitations of the proposal to allocate spectrum above 6 GHz would not affect the expected Bluetooth markets.

It is not improbable that regulators will revisit the concept of moving UWB above 6 GHz in the next 2–3 years. Once the UWB industry is able to build cost-effective radios in this spectrum and the application volumes of UWB are better understood, it will be easier for regulators to consider increasing the power to enable such a move without damaging the market viability of the technology. After several years of heated debate on UWB from a theoretical perspective, it is likely that the regulators will not be inclined to jump back into a new debate on UWB right away until a material body of measured data is available. The greater probability is that a few years will be allowed to pass before regulators will be ready to entertain the topic again.

10.13 Mitigation techniques

The second topic that has consistently proven to be confusing in both the UWB and regulatory community is the question of mitigation techniques. Mitigation techniques are methods by which the potentially harmful interference from UWB can be prevented or limited to an acceptable degree. For specific incumbent services, mitigation techniques can be used to gain additional protection against potential interference. In this area, there are two principal approaches being authorized in Europe; low-data-rate communications (LDC) and detect-and-avoid (DAA). In the USA, the 10-second rule was implemented as well as the prohibition on outdoor infrastructure. In Japan, a requirement was established that at least one device in a piconet must be connected to the power infrastructure (must be mains powered). The mitigation techniques described below are intentionally high-level. The object in this discussion is to give the reader a feel for the topic without providing overwhelming detail.

10.13.1 Low-data-rate communications (LDC)

Low-data-rate communications were envisioned as a means by which to address sensor-network applications of UWB. Sensor networks would include applications such as in a warehouse, where UWB tags are used to track the location of warehoused elements. An example of this would be when the military needs to set up a field warehouse. The food or ammunition has an attached UWB tag to allow soldiers to locate and distribute goods quickly.

A second, consumer, LDC example would be a network of intelligent thermostats, which are used to monitor and control the temperature of an office building. Ultra-wideband sensors built into the thermostats periodically communicate the temperature back to a controller, which then decides to turn on or off the air-conditioning unit accordingly.

In both of these cases, the amount of data being transferred is extremely low. Throughputs on the order of hundreds of kb/s are used. This compares with the hundreds of Mb/s or even Gb/s used to transfer files in HDR applications. Additionally, these sensors transfer data infrequently, to achieve a battery life which may be 1–3 years. Any mitigation technique designed to be used with sensor networks needed to preserve the battery life of the design.

If sensors did, in fact, transmit only infrequently, they would not cause enough interference to be of concern. However, as all technologies tend to evolve over time, there was a concern by regulators that the sensors would grow to have greater throughput and would become an interference source. If it were possible to guarantee that this did not occur, regulators would be much more comfortable with allowing higher power levels for the very brief periods when a sensor was on the air.

A sensor employing LDC limits itself to being on the air no more than 5% per second and 0.5% per hour. This timer mechanism requires very little sophistication to implement and therefore very little cost.

Low-data-rate communications also require almost no additional power to operate and so do not disturb the anticipated power budget of the device.

10.13.2 Detect and avoid (DAA)

In contrast to LDC is detect and avoid (DAA). The two mechanisms have been optimized for the environments in which they are expected to operate. Sensor applications have extremely low data communication requirements and very small power budgets. Low-data-rate communications are optimized to be used in this environment. Devices using DAA are optimized for extremely high-data-throughput requirements and are much less sensitive to power issues than a typical sensor application, but very sensitive to having their peak throughput restricted.

Conceptually, DAA is fairly straightforward. To use some specific frequencies, a UWB device must listen for operational incumbent systems. If an incumbent is detected, the UWB device must either modify its emissions to avoid interfering or it must abandon the spectrum in which the incumbent is operational. Separate DAA rules are being developed to cover protection of radar and fixed wireless-access systems.

In practice, this proves to be a difficult concept to implement. The first challenge is to distinguish incumbent transmissions from normal environmental noise. If recognition by the UWB device is not accurate, in the form of a false negative failing to identify an existing transmission, the UWB device may generate interference with the incumbent. If recognition by the UWB device fails in the form of a high false-positive rate, the UWB device operates sub-optimally. It avoids legitimately clear spectrum. As the false-positive rate increases, the value of DAA as a means of gaining access to spectrum for UWB implementers and their customers declines.

The second challenge that exists with DAA is that the detection mechanism needs to be fairly specific to the incumbent. As an example, in a fixed wireless-access (FWA) system, a cellular infrastructure is always present in most urban areas. If one were to detect and avoid based upon this downlink signal from the base station to the mobile device, the incumbent signal would be detected continuously. There would be no ability to share spectrum, regardless of whether there is a mobile device within interference range of the UWB system. Implementing DAA against a downlink signal would effectively cause UWB devices to abandon the shared spectrum in urban areas. Obviously, this defeats the point of sharing mechanisms. Instead, it is necessary to understand the FWA-system behaviour enough to know that the mobile elements in the network present periodic chatter. If a mobile device is in close proximity to a UWB system, detection can be based upon the mobile transmissions. Proximity to the mobile device can be estimated based upon the signal power and avoidance triggered when the mobile device is within interference range.

The third problem that must be overcome in DAA is avoidance. Once an incumbent has been correctly identified, it is necessary to modify the behaviour of the UWB device to avoid creating interference. The conceptually simple approach is to move to unoccupied spectrum whenever detection is triggered. However, if the UWB device is in the midst of communication, it would be necessary to coordinate any move between the two ends of the link such that the communication can be maintained in the new location. This is particularly true of streaming applications such as video, where lost packets will result in a visible or audible change for the consumer.

Before DAA systems can be deployed, agreement must be reached on how the effectiveness of a DAA system can be measured. The question is one of whether the DAA system sufficiently mitigates interference effects to the point that any interference is no longer harmful. In a radar system, this means that the probability of detection and identification of a target is not materially degraded. In an FWA system, UWB must not materially increase the number of outages experienced by users, must not decrease cell sizes and must not cause FWA customers to wait excessively to get access to the network.

In Europe, the regulatory bodies will be responsible for establishing the performance limits that DAA must achieve. But it will be ETSI that is responsible for describing how DAA performance is measured. As you will recall from the discussion of the EC's first mandate, the EC required that ETSI create any standards required in order to make UWB viable. To evaluate compliance to the regulations covering DAA, ETSI will need to create a measurement standard describing how conformance might be tested. Work on a measurement standard for DAA is expected to last until 2008 or later.

At the time of this writing, the particulars of DAA are still being actively debated in committee. However, the general outline of the process is now taking shape. In the case of DAA, the emitted power of the FWA system is known and can be used to estimate the distance between the FWA system and the UWB system. Depending upon the calculated distance, three zones are defined with different responses for each. Figure 10.3 provides a graphical summary of the DAA zoned approach used for fixed wireless access.

Assume for the moment that a UWB device powers up and listens for a nearby FWA transmitter. If one is heard and the distance between the devices is estimated to be within 36 cm, it is determined to be in zone one. In this case, the UWB device is required to cease transmission in-band and is required to suppress out-of-band emissions to a level of −80 dBm.

The emitted power of a FWA device at 36 cm is overwhelming. Turning off the transmission or moving to another band would be unavoidable. If desired, the UWB device may continue to test the spectrum. If the FWA emitter goes quiet or moves out of range, the UWB device may then resume transmitting.

If, when the UWB device listens for FWA transmissions, an interferer is determined to be operational within 20 m, this is considered to be a zone two scenario. The UWB device must cease in-band transmissions and suppress out-of-band emissions to a level of −70 dBm. As with zone one, the UWB device may resume transmitting once the FWA device leaves the area or ceases transmissions.

If a FWA device is detected at a distance beyond 20 m, this is zone three. Fixed wireless-access devices are unlikely to be affected by emissions from a UWB device at this distance. No response is required from the UWB system when a zone-three detection occurs.

Although the zoned system does an excellent job of avoiding interference to FWA devices that have become active, there was one problem scenario that remained unaddressed. If the UWB device were transmitting and a FWA device were to turn on, the signal from the

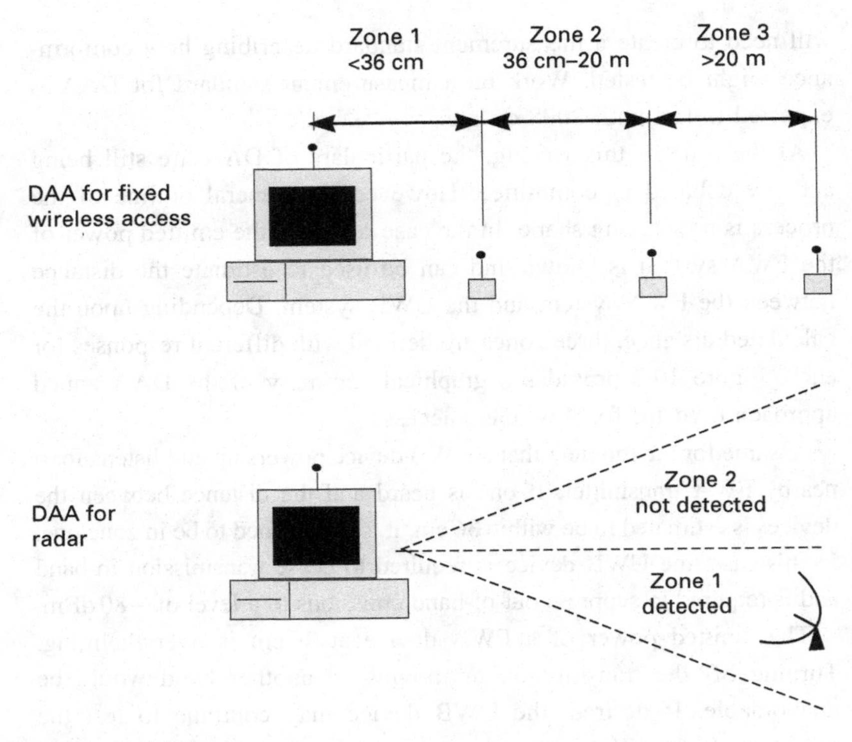

Figure 10.3 DAA zones

UWB system would be sufficient to mask the FWA base-station signal and the FWA device would be unable to register onto the network.

To deal with this scenario, it is necessary to have periodic silence periods in the UWB transmission. These silence periods need to be long enough to allow the FWA device to register, but not so long that the value in sharing the spectrum is destroyed. The exact length of these silence periods is still being debated.

10.13.3 Ten-second rule

The third mitigation technique that will be discussed is called the ten-second rule. This is a concept that was originally developed by the FCC. The concern that was expressed during initial proceedings was that UWB devices would be in a continuous search mode looking for new communication partners. This act, if repeated often enough and long enough, has the potential to account for a significant volume of activity and therefore interference potential.

Somewhat more importantly, it would generate activity both indoors and outdoors. Indoor operations are generally considered to be somewhat safer, because the walls and ceiling of a structure attenuate the strength of the signal significantly.

To mitigate this risk, the FCC required that any device which was unable to find a communications partner after 10 seconds of beaconing was required to go quiet. The FCC rule is silent on the question of what conditions, if any, would allow the device to resume searching for a partner. It is generally believed that a user-initiated event which begins a new search would be permissible, for instance, a power-cycle or a transfer-start command, but this has not been tested.

While this rule is not mandated in every country, it is generally expected that manufacturers will incorporate it into their offerings worldwide, regardless of that fact. This feature does not materially affect any identified user scenario and does not cost the manufacturer anything to implement.

10.13.4 No outdoor infrastructure

Another invention of the FCC was the prohibition on outdoor infrastructure. As with the 10-second rule, the FCC was concerned that outdoor use of UWB had a disproportionate ability to generate interference. Outdoor infrastructure is not closely defined by the FCC in its rulings. One could reasonably question the definition of 'outdoor' and 'infrastructure' as well, for that matter. Conversations with well placed individuals in the FCC suggests that the FCC will step in to define the terms more closely only if it appears that companies are attempting to circumvent the intent. If, for instance, a kiosk network were deployed wherein each had a rain cover and open sides, it is unlikely that the FCC would consider this favourably.

While LDC and HDR flavours of UWB are fairly unlikely to be deployed in a manner that would justify this technique, imaging uses of

UWB could potentially be very problematic. Imagine a system wherein UWB is used for its imaging properties to create an electronic fence around a property. In this case, the UWB signal would need to be on constantly, or at least a high percentage of the time, to provide reliable detection. Likewise, this type of application would be used outdoors almost exclusively. From an interference perspective, it would be the worst of all combinations.

10.13.5 Mains attached

Just as the FCC was concerned by outdoor uses of UWB, the Japanese government also chose to take actions to prevent outdoor uses of UWB. Most notably, the Japanese chose to require at least one UWB device in a piconet to be attached to a power outlet or to be 'mains powered'. The belief behind this requirement was that the size of a UWB piconet was small. If one device were required to be attached to a power socket, it would effectively restrict UWB to indoor uses.

This is one of the more controversial mitigation techniques that has been suggested. This requirement has the potential to limit many peer-to-peer applications, such as transfers between mobile devices, as well as automotive applications.

The work in the ITU and Europe on outdoor activity factors would suggest that this type of restriction is potentially unnecessary. Interference potential in HDR UWB is highly correlated with the use of streaming video. Because streaming video requires an infrastructure to deliver it, and this infrastructure is overwhelmingly available indoors, it is generally believed in the UWB community that a restriction on outdoor infrastructure will be more effective at allowing appropriate operations while providing reasonable protection to incumbents. It is likely that further conversation on this topic will be conducted when more data are available.

10.14 Summary

Radio regulation tends to be a rather specialized area of expertise which most engineers have only a limited need to understand. Because UWB is not only a new technology but also a new regulatory paradigm, it is necessary to establish a basic understanding of regulatory processes, intentions and actions in the administrations most likely to affect policy worldwide.

The FCC in the USA principally play a role of chief experimenter. They have the resources to examine new concepts and the innovative fervour required to issue rules in support of new technologies. The European contribution is principally one of consensus. The processes that have been constructed have the ability to gain alignment of a large number of administrations, which seed international discussion and so act as a major influence to shape world opinion. Because of this, understanding the European processes surrounding regulations in general and UWB in particular becomes important.

Ultimately, what becomes most important in establishing policy is the give and take between forces promoting protection of existing services opposing forces that are developing new services. In almost all cases, the national regulatory stance is principally established by law. Within that legal charter, regulators have to decide which changes can be supported and which must be opposed or modified. Industry groups acting in support of their own commercial interests then attempt to persuade and compromise their way to a conclusion. In the discussions on UWB, this has been a particularly challenging effort because of the very large number of vested interests that must be represented. Gaining agreement among all of them is extremely challenging.

Ultimately though, compromises and agreements are being made. Concerns about aggregation of interference among low-data-rate devices were solved by the development of LDC rules. Detect and avoid is being developed to answer concerns about HDR devices operating in close proximity to radar and fixed wireless access. The phased approach was agreed to minimize grey-market potential by enabling manufacturers to sell products today. There is still a great deal of work to do to change the UWB regulatory experiment into a polished system.

References

[1] Federal Register, Vol. 63, #182, Sept 21, 1998, Page 50184, See www.fcc.gov/Bureaus/Engineering_Technology/Documents/fedreg/63/50184.pdf, accessed 4 October 2007.

[2] http://circa.europa.eu/Public/irc/infso/radiospectrum/library?l=/public_documents_2005/rsc12_june_2005/second_mandatepdf/_EN_1.0_&a=d, accessed 10 March 2008.

[3] www.ofcom.org.uk/research/technology/archive/cet/uwb/uwbpans/, accessed 4 October 2007.

[4] http://eur-lex.europa.eu/LexUriServ/site/en/oj/2007/l_055/l_05520070223en00330036.pdf, accessed 4 October 2007.

11 Tragedy of the commons

When a property is owned by a single person or entity, it is the responsibility of that person to maintain the property in good condition. If he or she fails to do so, it is entirely to his or her own disadvantage. The value of the property will decline. Likewise, any profit that may be derived is solely to the benefit of the property owner. The owner is therefore motivated to manage the property in a manner that optimizes its profitability.

If a property is communally owned and the individuals are not closely associated with one another, the incentives change. The individuals involved will each attempt to maximize their own personal profit from the property while attempting to minimize their contribution to the upkeep and maintenance. The property becomes overused, while at the same time maintenance is neglected. The value of the property goes into a downward spiral. This behaviour is known in regulatory circles as 'the tragedy of the commons' and is frequently used to describe potential risks of unlicensed spectrum. More specifically, this describes spectrum wherein anybody is able to build devices which operate without the enforcement mechanisms that manage use.

11.1 Ultra-wideband spectrum saturation

Ultra-wideband is an unlicensed spectrum of this type. Ultra-wideband is somewhat different in considering spectrum saturation than longer-range technologies. The fact that its power is so strictly limited reduces the number of devices that might cause interference to UWB as well as the number of devices that it might interfere. This creates a form of geographic limitation (de facto licensing of a sort), even if it is not a statutory restriction applied by regulators.

There is one situation where spectral saturation is a reasonable potential. In densely packed office environments, if UWB were set up for wireless-monitor or wireless-docking applications (which include wireless-monitor functions plus peripheral connectivity), it would be possible to saturate the available spectrum. A lot of the risk depends upon the specific implementation.

If one attempts to build the wireless monitor function with the objective of maximizing spectrum utilization, there are several options. First, computer monitors, unlike televisions, are rarely the target of streaming-video content. The video content on monitors is very sporadic in nature. It normally changes as a result of an event generated by the user. As an example, the user enters a new address in a browser and the screen is refreshed. Until the user finishes consuming what is on the screen, further transmission isn't necessary. If the implementer takes advantage of this fact by storing the most recent screen update on the monitor end, it is not necessary to continue sending the same image across the link to keep the screen refreshed. This is a simple form of compression that would be effective at reducing bandwidth consumption. Taken further, some compression algorithms, such as JPEG2000, have the ability to do various levels of compression. It would be possible to implement a wireless-monitor algorithm such that a moderate amount of compression were implemented that did not tax the processing resources to the degree that MPEG 2 or 4 would.

By contrast, if an implementer were to choose to use uncompressed video and maximize the resolution and size of the display with the emphasis on visual quality and low cost, the situation would be entirely different. Under this scenario, the implementer would likely perform channel bonding. That is, they would attempt to transmit on multiple UWB channels simultaneously to get the effect of a much wider-band radio. In this situation, the spectrum is consumed very rapidly. It would be unlikely that such a system would be capable of existing in an office environment (with cubicles or open bull-pits). Saturation would almost inevitably occur.

If saturation were to occur in this manner, the results would probably not be that severe. A corporate directive regarding the choice of wireless monitors would probably be sufficient to resolve the matter. Economic impact would be limited to the company or companies producing the spectrum-consuming wireless-monitor system.

11.2 Saturation of WLAN due to PAN applications

The second area of concern related to spectrum saturation, which is worth discussion here, is not the excessive and uncoordinated use of UWB, but is instead the potential for platform integrators to create a tragedy of the commons effect in 2.4 GHz and 5 GHz WLAN spectrum by attempting to move PAN applications onto WLAN radios to save the cost of an additional radio.

This is a somewhat complex series of relationships that require a little bit of explanation. To begin, of the three markets active in the PAN and LAN spaces (PC, CE and mobile handset), only the mobile-handset groups design radio systems for a living. The PC and CE system integrators tend to purchase radio subsystems, which they incorporate into the larger design. Their need to understand the radio and, therefore, their depth of expertise in most cases is rather limited. They tend to view their system as being a PC or CE component with connections for communication paths. Their system-analysis responsibilities are limited to the outside boundaries of the box. What goes on in the communications network is overwhelmingly the responsibility of a communications-service provider, such as the telephone company or the cable-television provider.

Within the box, they have been trained to optimize cost by reducing duplication wherever possible. They are conditioned to be minimalists. They combine the functionality of two devices into one. They find new ways to share functionality instead of repeating it. This approach works well for data-communications environments. It even works reasonably well with systems containing one radio or two radios whose functionality is separated by a substantial performance gap such that radio aggregation is not tempting.

When PC and CE manufacturers attempt to optimize the number of radios on a platform and the radio performance is not obviously differentiated (or the differentiation is minimized), they attempt to use a single radio instead of two or more. Specifically, they find a strong inclination to employ a single WLAN radio instead of a WLAN and a PAN radio. If successful, this will allow them to save the cost of the PAN radio entirely and thereby reduce the platform price.

Supporting this inclination to merge radios is the natural competition among radio-chip manufacturers. Each radio manufacturer has a financial incentive to employ its own device wherever it might fit in order to maximize sales potential. The old saying, 'If you only have a hammer, all your problems are nails,' comes to mind. The urge to reduce costs and the urge to sell into adjacent markets become mutually supporting.

For this scenario to work out for the customer, two assumptions must be valid. First, it must be true that the spectrum available is sufficient to support all of the applications that will be run on the platform. It must also be true that if the spectrum proves to be insufficient, that saturation is a 'soft' obstacle. The functional impact on the customer must be relatively minor and the radio can be modified after the point that saturation is encountered to correct for the problem.

First, consider the question of sufficiency. In work done on UWB in Europe and in the ITU, it was found that a single MPEG 2 video stream transfers approximately 19 times as much data as is transferred in a similar period by all WLAN applications used today combined. In a home environment, credible sources are estimating that three or four high-definition (MPEG 2 or 4) streams will be needed in the USA and somewhat fewer than that in other markets. Conservatively, this would suggest that the support of two or three concurrent video streams would represent about 30–57 times the traffic volume than what is currently being carried by WLAN networks.

Spectrum in 2.4 GHz and 5 GHz ISM bands used for LAN traffic also contains a number of users other than WLAN devices. Cordless phones exist in both bands. Microwave radios generate unintentional interference in the 2.4 GHz band. In the USA, ham television operators have priority in the 2.4 GHz bands. Even older versions of WLANs have to be taken into consideration, as they will continue to consume spectrum.

Unfortunately, there is no quantitative data available to suggest the point at which saturation of the ISM spectrum will occur. Anecdotally, there is some belief that the 2.4 GHz band is already fairly crowded. However, without an analysis, it would seem that any effort to increase bandwidth used by this volume should be considered carefully. The assumption of sufficient spectrum seems speculative at best. The second assumption about saturation being a soft obstacle is also questionable. Having encountered a saturation condition once in wireless local loop, allow us to offer this model of how saturation is likely to proceed.

During the growth phase of the technology, radios are developed and deployed that do a superb job at addressing consumer demand. Positive word of mouth is generated, causing increased sales. Multiple manufacturers enter the market. The increased volume of manufacturers and the maturing of the technology cause prices to fall, products to become integrated into larger systems and volumes to rise further.

All of these positive market events cause spectrum use to increase. The first sign of difficulty occurs in a dense urban area such as Paris (82 000 people/km^2). People purchasing new systems begin to return them because they are unable to get a reliable link in place. Streaming-video applications experience failure first. The streaming nature of the video does not allow for retransmission and so interference shows up as visual artefacts.

Return rates to distribution channels and to video manufacturers begin to rise. These organizations have a normal return rate of approximately 3–6%. Consumers are not simply returning $5 radios. They are probably returning televisions that may be several hundred dollars, because of the $5 radio. This creates a financial amplification effect on the integrators. When acceptable return levels are exceeded, manufacturers are forced to pull models with integrated radios. Distribution channels begin to advise their customers not to buy radio-enabled devices. Reviewers begin to comment negatively in magazines, and blogs. The word-of-mouth inertia that was responsible for driving the growth of the market now drives its decline.

Once the market has experienced the saturation of the spectrum and collapse of the sector, it becomes difficult to convince manufacturers to risk taking the chance again. Unlike the localized PAN saturation described earlier, which affected only a small number of market participants, this type of saturation reaches much further and is much harder to correct. If manufacturers do choose to return to the spectrum, both the customers and the distribution channel have to be convinced that the problem is no longer present and that saturation will not recur. This is a challenging and expensive task that takes years to accomplish. If the integrators are more knowledgeable about the cost–spectrum trade-offs, saturation issues can be avoided.

Each platform should be designed with enough radios (radio spectrum) to support the applications envisioned for that platform. High bandwidth, short-range applications and applications targeting mobile platforms should be concentrated on PAN radios, which were designed for that purpose. Longer-range applications, which need to penetrate walls, should be reserved exclusively for WLAN radio architectures.

If one looks at how mobile-phone carriers use their spectrum, they estimate the total traffic that is supportable and the acceptable performance metrics that consumers will experience. They operate up to those levels. Obviously, in unlicensed spectrum this is harder to do.

Today, the number of devices needing multiple radios is relatively small. Mostly, PCs are affected. However, the cost and size of these radios is shrinking rapidly at the same time that intelligence and storage capacity on portable platforms is growing. The need for multiple radios is imminent for set-top boxes, mobile phones, portable PCs, cameras and a host of small battery-powered devices. While the situation is relatively manageable today, the problem is expected to escalate over time.

Manufacturers must estimate these factors for the industry as a whole and not just for themselves. Simulations should be run to estimate the probable saturation conditions. This information should then be used to determine which radios are present on the platform. By spreading the spectrum load among several radios, the likelihood and severity of any saturation condition is materially reduced.

11.3 Summary

The concept of ‘tragedy of the commons’ is a well known condition, which applies to the use of unlicensed spectrum. Since UWB falls under this category, it is incumbent upon manufacturers and integrators to be aware of these situations and to take reasonable steps to avoid this type of failure. In the most direct interpretation of the condition, PAN applications saturate UWB spectrum. This condition is not overly concerning because the limited range of UWB acts as a mitigating factor to limit the number of participants in any interference scenario and to limit the affected number of players if saturation does occur.

What is probably more concerning is a system issue. Manufacturers of PCs, STBs and a host of battery-powered devices have the potential to have both WLAN and PAN radios included. There are forces which make it appear reasonable to assign PAN applications, such as docking and wireless video, to WLAN radio technologies to save the cost of the additional radio. If done, this has the potential to saturate the spectrum in the 2.4 GHz and 5 GHz WLAN frequencies.

If one approaches the problem from a system perspective, the correct answer is to assess the spectrum needs of all applications supported by a platform. Sufficient bandwidth should be present on the platform to service those applications appropriately, regardless of the number of radios required to perform the task. If this is done, the likelihood of saturation issues diminishes sharply and the severity of the issue if it occurs may be reduced.

Appendix: Reference documents

The documents referred to below are original source materials, which can provide substantially more detailed information about the technical, procedural and historical aspects of UWB. The specific technical analysis that was done as part of the regulatory and standards processes is not referenced here. It is overwhelming in its volume after almost nine years of active debate on the topic. If one wishes to search for this level of material, it would be necessary to gain access to the specific FCC comments made during their various hearings, the submissions made during the TG3 meetings (a CEPT/ECC subcommittee) as well as other subcommittees and the IEEE 802.15.3a committee presentations. One should generally be cautious about taking these documents on face value. It was not uncommon for competitive fervour to be expressed as often as technical rigour. Most, if not all, of the participants were afflicted with this condition at one time or another.

Regulations (FCC)

(1) *Notice of proposed rulemaking* – This is the FCC's initial announcement indicating an intent to develop rules on UWB. It essentially provides the first statement about policy. It is followed by comments from industry sources.

www.fcc.gov/Bureaus/Engineering_Technology/Notices/2000/fcc00163.doc (accessed 26 December 2007)

(2) *Comments* – More than 700 comments were received to the initial FCC notice of proposed rulemaking issued in 2000. In this speech before the house subcommittee on telecommunications and the Internet, Julius Knapp talks about UWB.

Testimony of Julius Knapp before the US House of Representatives Committee on Energy and Commerce

Subcommittee on Telecommunications and the Internet, 5 July, 2002, p. 5.
www.fcc.gov/Speeches/misc/statements/knapp060502.pdf (accessed 26 December 2007)

(3) *FCC first report and order* – This is the first FCC Report and Order authorizing ultra-wideband for use in the United States.
www.fcc.gov/Bureaus/Engineering_Technology/News_Releases/2002/nret0203.html (accessed 26 December 2007)

(4) *Additional FCC Actions* – All of the FCC rulings related to UWB are kept at:
http://wireless.fcc.gov/spectrum/proceeding_details.htm?proid=288 (accessed 26 December 2007)

(5) *NTIA papers related to UWB* – As part of the US proceedings on UWB, the NTIA analyzed the interference effects as they related to government spectrum holders. These papers describe those results.
http://www.ntia.doc.gov/osmhome/uwbtestplan/index.html (accessed 26 December 2007)

(6) *Multispectral Solutions Inc. archives* – MSSI was an early pioneer in UWB technologies and developed location devices and covert-communication devices for the military. They were active in commenting on the FCC processes and kept extensive files on the proceedings.
http://www.multispectral.com (accessed 26 December 2007)

Regulations (European)

(7) *First European Mandate* – This mandate was generated in cooperation between the European Commission and the Radio Spectrum Committee. It directs CEPT and ETSI to develop standards necessary to bring UWB to market. The ECC, acting as an engineering arm of CEPT, is engaged as well.
http://circa.europa.eu/Public/irc/infso/radiospectrum/library?l=lpublicsdocumentss2004/rsc7/rscom0408smandatesonsuwb/_EN_1.0&a=d (accessed 26 December 2007)

(8) *Report 64* – This document is the result of the first mandate. It codifies the logic of zero interference. It is skewed heavily to a

position of protection for incumbent services. This fact is commented upon in the EC's second mandate document. While this document does characterize the state of discussion in the working committees, caution is advised in taking the analysis too much at face value. There are significant internal inconsistencies. It would be best to view this document as a statement of the incumbent service providers' position.

http://www.ero.dk/doc98/Official/pdf/ECCREP064.pdf (accessed 26 December 2007)

(9) *Second European Mandate* – The second mandate provides additional direction to CEPT/ECC. The EC attempts to rebalance the protection vs. innovation weighting that has occurred in first mandate work. The compromise that came out of this work is principally reflected in the EC decision. Note the difference between Report 64 recommendations, the EC decision and the FCC decision.

http://ec.europa.eu/information_society/policy/radio_spectrum/docs/by_topics/final_second_uwb_mandate.pdf (accessed 26 December 2007)

(10) *Third European Mandate* – The third mandate authorizes clean-up work. As a result of the second mandate, several recommendations are made by the technical committees, which would have been challenging to place into legislation or contain conclusions that are unsupported by the work performed. In this regard, it is still possible to see a strong leaning toward incumbent interests in the work. This document authorizes work to correct those issues.

http://ec.europa.eu/information_society/policy/radio_spectrum/docs/current/mandates/3_ec_to_cept_uwb_06_06.pdf (accessed 26 December 2007)

(11) *EC Decision on UWB* – This is the initial decision allocating spectrum for UWB in Europe. It was the starting point for decisions reached in a number of other countries as well.

http://ec.europa.eu/information_society/policy/radio_spectrum/docs/ref_docs/rsc21_public_docs/rscom07_60draft_spec_uwb.pdf (accessed 26 December 2007)

(12) *R&TTE Directive* – This document is the legal foundation on which European regulation is principally conducted under the current EC structure. This is necessary background reading to understand the philosophical position of the EC in pursuing new technology as well as authorization for some of the structures such as the relationship between the EC and RSC.
http://ec.europa.eu/enterprise/rtte/dir99-5.htm (accessed 26 December 2007)

(13) *Mason Report* – This was the principal economic assessment(s) of the likely impact of UWB. It was conducted by Ofcom (UK regulatory agency) and is focused on the effects in England. The report was conducted in two major stages.
www.ofcom.org.uk/research/technology/archive/cet/uwb/uwbpans/uwmpans.pdf (accessed 26 December 2007)
www.ofcom.org.uk/research/technology/archive/cet/uwb/background_uwb_rpt/summary_uwb (accessed 26 December 2007)
www.ofcom.org.uk/research/technology/archive/cet/uwb/background_uwb_rpt/tech_evaluation (accessed 26 December 2007)

Standards

(14) *ECMA 368* – This is a copy of the standard covering the UWB PHY and MAC, which is used by the WiMedia Alliance, the Bluetooth SIG and the USB IF.
www.ecma-international.org/publications/standards/Ecma-368.htm (accessed 26 December 2007)

(15) *ECMA 369* – This is a copy of the standard covering the PHY–MAC interface. If a manufacturer decides to implement a PHY and a MAC as separate physical components, this describes how they connect to each other. If a manufacturer chooses to implement an integrated solution, this is not relevant. When the market opened, several manufacturers employed this specification. As the market evolves into single-chip solutions, it will become increasingly irrelevant.
www.ecma-international.org/publications/standards/Ecma-369.htm (accessed 26 December 2007)

(16) *ISO/IEC 26907* – This is the ISO approved version of the ECMA 368 standard.
www.iso.org/iso/iso_/catalogue/catelogue_tc/catalogue_detail.htm?csnumber=43900 (accessed 26 December 2007)

(17) *ISO/IEC 26908* – This is the ISO approved version of the ECMA 369 standard.
www.iso.org/iso/iso_catalogue/catlogue_tc/catalogue_detail.htm?=43901 (accessed 26 December 2007)

(18) *WTO trade agreement covering standards* – The full text under 'Technical barriers to trade'. This is not intended to be an exhaustive covering of the topic. It is more of an insertion point for somebody who wishes to investigate the topic more fully.
www.wto.org/english/docs_e/legal_e/17-tbt.pdf (accessed 26 December 2007)

(19) *WTO introduction and discussion of technical barriers to trade* – As above, this is intended as an overview of the topic of trade barriers.
www.wto.org/english/tratop_e/tbt_e/tbt_info_e.htm (accessed 26 December 2007)

(20) *IEEE 802.15.3a* – The IEEE 802.15.3a committee was the first standards body to undertake development of the UWB MAC and PHY. They worked for approximately three years but were ultimately unsuccessful at reaching agreement and ceased operations. The contributions to the committee tracked some of the early evolution of UWB modulation techniques and expended a great deal of effort specifically in comparison of DS vs. MB-OFDM flavours. By the time that the work began, PPM was already in disfavour and so is not included in any of this material.
http://grouper.ieee.org/groups/802/15/pub/Download.html (accessed 26 December 2007)

Special interest groups

(21) *WiMedia Alliance* – This is the home page for the WiMedia Alliance. It gives data about membership, policies and activities. The documents describing bylaws and intellectual property are

publicly available. A members-only portion of the site contains documents (current and archived).

www.wimedia.org (accessed 26 December 2007)

(22) *Bluetooth Special Interest Group* – This is the home page for the Bluetooth Special Interest Group. It gives data about membership, policies and activities. The documents describing bylaws and intellectual property are publicly available. A members-only portion of the site contains documents.

www.Bluetooth.org (accessed 26 December 2007)

(23) *USB Implementer's Forum (responsible for Wireless USB)* – It gives data about membership, policies and activities. The documents describing bylaws and intellectual property are publicly available. A members-only portion of the site contains documents

www.usb.org (accessed 26 December 2007)

(24) *1394 Trade Association* – It gives data about membership, policies and activities. The documents describing bylaws and intellectual property are publicly available. A members-only portion of the site contains documents.

http://1394ta.org (accessed 26 December 2007)

(25) *ZigBee Alliance* – It gives data about membership, policies and activities. The documents describing bylaws and intellectual property are publicly available. A members-only portion of the site contains documents. The connections of Zigbee to UWB are somewhat tangential. It is included here for the benefit of the excessively curious.

http://zigbee.org (accessed 26 December 2007)

Author biographies

Stephen Wood began working on ultra-wideband in 2001 about eight months before the first FCC ruling. He was one of the original founders of the OFDM Alliance, which advocated the multiband OFDM modulation in the IEEE work and eventually merged into the WiMedia Alliance. In WiMedia, Mr Wood was elected to the role of President. It is a role that he has maintained for the last three years.

Mr Wood is involved extensively in structuring WiMedia, including its intellectual-property rules, relationships with other SIGs and relationships with standards organizations. He has also been very active in regulatory development in the USA, Europe and the International Telecommunications Union.

His professional background includes approximately 25 years in mainframe communications, local-area networking and wireless local area networking, cellular communications and personal-area networking. His responsibilities include product marketing and strategic marketing roles.

Dr Roberto Aiello is the founding CEO and now CTO of Staccato Communications. Prior to working at Staccato, he was founder, President and CEO of Fantasma Networks, an ultra-wideband product company. Previously, Dr Aiello led the wireless research and built the first documented UWB network at Interval Research, Paul Allen's research laboratory. Earlier, he held senior positions at the Stanford Linear Accelerator Center and the National Superconducting Super Collider Laboratory in Texas.

Dr Aiello is a recognized leader in the UWB community, and his efforts were instrumental in getting UWB spectrum allocated in the United States. Dr Aiello holds a PhD in physics from the University of Trieste. He serves on several advisory boards and is the author of more than 20 patents on UWB technology. This is his second book on UWB.

Index

5 GHz 190
6 GHz 175, 176
75% approval 112
802.11 23
802.11a 19
802.15.3a 50, 110, 112
1394 Trade Association 97, 199
26908 105

abbreviated data frames 89
activity analysis 159, 163
activity factor analysis 160
adaptation Layer (PAL) 61
addressing 48
ad-hoc networking 87
adopter 133, 134
adopters 116, 135
aggregate interference 162
aggregation 170
aircraft 169
analog-to-digital converter (ADC) 71
anchor 57, 91
antennas 32, 73
associates 135
association 99
association frames 89
authentication 90
automobiles 169
automotive radar 7
automotive UWB 2

backbone 141
ball grid array (BGA) 72
bands 38, 41
beacon 53
BiCMOS 71
binary phase-shift keying (BPSK) 42
Bluetooth 3.0 41, 61, 92
Bluetooth 81, 92, 144
Bluetooth Special Interest Group
27, 92, 116, 117, 120, 125, 176, 199
brand 143, 145
bridge function 88

cable 99
cable replacement 83, 93, 122
CEPT (European Conference of Postal and Telecommunications Administrations or Conference Européenne des Administrations des Postes et de Télécommunications) 154
CEPT 154
certification 119, 131
certification test suit 132
Certified Wireless USB (CWUSB or WUSB) 27, 41, 82, 123
channel bonding 138
characterizing 159
China 107
Class 1 radios 92
Class 2 radios 92
Class 3 radios 92
clusters 84
CMOS 70–2, 76
co-exist 66
common radio platform 37, 51, 52, 89, 91, 117, 120, 124
consensus 152, 153
consensus model 152
consolidation 143
contention 49
contributor 133, 134
control frames 89
converging the WAN, LAN and PAN 139
core specification 95
correcting these errors 48
cost 26
CWUSB 61

DAA 53, 171
daisy chain 97
detect and avoid (DAA) 148, 167, 177, 179
digital video recorders (DVRs) 34
direct sequence 15, 112
directional 74

display port 21
distributed reservation protocol 58, 60
Douglas Adams 81
DS 124, 198

ease of use 86
EC 156, 157, 158, 164, 168
EC decision 196
ECC (Electronic Communications Committee) 155
ECC 158
ECMA 368 48, 103, 197
ECMA 369 13, 197
Ecma International 37, 103, 113, 133
ECMA-368 37
economic impact assessment 167
ETSI 105, 106, 107, 158, 180
European Commission (EC) 105, 107, 155
European Standards and Technical Institute (ETSI) 103
European Union 155
evolve 137
extended board 133

fairness 62
FCC 12, 16, 107
Federal Communication Commission 13
file transfer 4, 19, 20
FireWire 97
First European Mandate 195
First Mandate 158, 163
First Report & Order 195
fixed frequency interleaving (FFI) 45
fixed outdoor antennas 160
fixed wireless access (FWA) 166, 179
frequency reuse 33

game consoles 9
grey market 167
ground-penetrating radar (GPR) 2, 7

hard reservation 60
harmful interference 68, 161, 162, 177
HDMI 21
HDR 4, 5
hibernate 56
high data rate (HDR) 3
home networking 140
host–controller interface 95
hot spot 140, 141

IEEE 110
IEEE 15.3a 15, 104, 197
IEEE 802 15.4 124
imaging 7
indoor 160, 173
inductance 72
induction 100
information exchange (IE) 55
innovation 152
integrated 70
integration 66, 70, 76, 79
intellectual property 115
intellectual-property rights 120, 125
interference 150, 159
International Standards Organization 103
International Telecommunications Union (ITU) 152
Internet protocol (IP) 61
interoperability 66, 95, 115, 118, 119, 131
in-wall imaging 7
ISO 108, 109
ISO/IEC 26907 105, 198
isolation 73
ITU 153, 190

joules per bit 30
JPEG2000 188

kiosk 10, 138
kiosk networks 6

L2CAP 94
LAN 33
land grid array (LGA) 72
latency 62
LDC 178
LDR 6
licensing 149, 158
licensing fees 108
location 28
location tags 28
logo 131
low-data-rate communications (LDC) 177
low-temperature co-fired ceramics (LTCCs) 72

"magic" $5 price point 142
mains powered 91, 184
MAS 60
Mason report
master-slave 94
matching networks 75
MB-OFDM

media-access control (MAC) 48, 103
media allocation slots (MAS) 54
mesh 31
mesh networking 89
Metcalf's Law 144
microscheduled management command (MMC) 64
military 11, 28
minimally coded video 148, 175
minimally compressed video 172
mitigates 67, 177, 180
mitigation techniques 166, 175
MPEG 21
MPEG 2 160, 173, 188, 190
MPEG 4 163, 173
Multi-Band OFDM Alliance 50
Multi-Banded Orthogonal Frequency Division Multiplex (MB-OFDM) 15, 112
multicast 60
multipath 42
multiple radios 76
mutually assured destruction (MAD) 128

National Administration 154
near-field communications NFC 99, 100, 103
neighbours 55, 60
network allocation vector 59
noise 73
Notice of Proposed Rulemaking (NPRM) 16
NTIA 195

OFCOM 168, 197
OFDM 38, 74
omni-directional 75
one company, one vote 113
one man, one vote 113
outdoor infrastructure 183
overhead 85, 87

packages 72
packaging 66, 70, 72
PAN 10, 32, 120, 122
PANs 32
peer-to-peer 87, 88, 97
period start time (BPST) 54
personal-area network (PAN) 87, 88, 110
personal video recorders 140
phased approach 166, 167
physical layer (PHY) 37, 103
PIN (personal information number) 99, 100
power 30
power efficient 139
power management 85, 90
power requirements 137
PPM 14, 15
prioritized contention access 58
private reservation 63
profile specifications 95
promoter 116, 132, 134, 135
proposals 111
protect 152
protection 154
protection zones 167
protocol adaptation layer 89
pulse position modulation (PPM) 13

QoS 91
quality of service 90

R&TTE Directive 197
radar 167, 179
radio astronomy services (RAS) 154, 167, 171
radio integration 139
Radio Spectrum Committee (RSC) 155, 156
RaNDZ 126
reasonable and non-discriminatory (RaND) 126
reasonable and non-discriminatory with zero royalty 126
report 64 163
report and order 17
reservations 51
royalties 108, 126, 128

safe medium access slot(s) rule 62
satellites
Second European Mandate 165, 196
secure 82
security 85, 86, 89, 95, 122
Security Mode 1 96
Security Mode 2 96
Security Mode 3 96
security perimeters 7
sensor networks 6, 124, 178
set-top boxes (STB) 34, 140
SiGe 76
silicon–germanium (SiGe) 70
single entry 170
single entry interference 161
single entry scenario 166
smart antennas 138
soft reservation 60
special-interest groups 115

specmanship 23–4
spectrum regulation 149
spectrum saturation 146, 187, 189
spectrum underlay 148, 150
speed 19
standard data frames 89
standards 103
STB 10
STBs 9
streaming audio 5
streaming video 4, 5, 21, 34, 58
subcarriers 42
superframe 54

tagging 28
Task Group 3 (TG3) 157
Technical Group 3 (TG3) 159
ten-second rule 182
test labs 119
TG20 113
TG3 163
Third European Mandate 168, 196
throughput 25
through-wall imaging 7
time-frequency code 45
time–frequency interleaving (TFI) 45
trade agreements 104
tragedy of the commons 187
transmit frequency code (TFC) 52
transmit power control 56
transport specifications 95
troll 129

uncoded 148, 175
uncompressed 172
uncompressed video 21, 22, 172
underlay 2, 2–3, 16
unintentional emissions 151
universal serial bus (USB) 27
universal serial bus Implementer's Forum 121
unusued DRP reservation announcement 60
urban area 191
USB 81, 144
USB addressing 63
USB IF 117, 125, 134
USB Implementer's Forum
UWB Forum 124

video 160

wide-area network (WAN) 9, 31
WiFi 87, 116, 146
WiMedia 37, 144
WiMedia Alliance 84, 106, 117, 125
WiMedia Layer 2 Protocol (WLP) 81, 86
WiMedia MAC 51
Wireless 1394 61, 81, 97
wireless docking 145
wireless USB 81
WLAN 19
WLP bridging 89
World Trade Organization (WTO) 110
WUSB host channel 64
WUSB MAC 51, 63

zero interference 164
Zigbee Alliance 81, 124
zone one 181
zone three 181
zone two 181

Printed in the United States
by Baker & Taylor Publisher Services